"As a mayor who worked to reduce the carbon emissions of our city, I counted on our citizen leaders, parents, and children to help us develop our city's Climate and Energy Plan and one of the strongest climate ordinances in the nation. Mary DeMocker effectively blends her creativity, scientific knowledge, and commitment to our planet's well-being to support and encourage positive activism. She knows that it will take all of us and she shares my optimism that we can make a difference in our homes, our cities, and our nation. Her book makes it easy for us to create a better future for all children."

— **Kitty Piercy,** three-term mayor of Eugene, Oregon, voted "Most Valuable Local Official" in the U.S. by the *Nation* in 2010

"DeMocker brilliantly links parenting skills that we already have to the social and political changes that we can make — and are making — to stabilize the Earth's climate for future generations. This is a guilt-free roadmap to saving ourselves in order to save the planet."

— **Stephanie LeMenager,** professor of environmental studies at the University of Oregon and author of *Living Oil*

D0176595

THE PARENTS' GUIDE TO
CLIMATE
REVOLUTION

THE PARENTS' GUIDE TO
CLIMATE
REVOLUTION

100 Ways to Build a Fossil-Free Future,
Raise Empowered Kids,
and Still Get a Good Night's Sleep

MARY DeMOCKER

Foreword by Bill McKibben

New World Library
Novato, California

New World Library
14 Pamaron Way
Novato, California 94949

Illustrated by Hannah Peck
Text design by Tona Pearce Myers

Chapter 20 appeared in a different form in *ISLE* journal and an earlier version of chapter 98 appeared in *The Oregonian*.

Library of Congress Cataloging-in-Publication Data is available.

First printing, April 2018
ISBN 978-1-60868-481-6
Ebook ISBN 978-1-60868-482-3
Printed in Canada on 100% postconsumer-waste recycled paper

 New World Library is proud to be a Gold Certified Environmentally Responsible Publisher. Publisher certification awarded by Green Press Initiative. www.greenpressinitiative.org

10   9   8   7   6   5   4   3   2   1

For Zannie and Forrest,
with all my love

It's not either/or, parent or save the Earth. The question is how your decisions as a parent can teach your children to love the thriving planet, and to understand that love is a way of acting in the world.
— KATHLEEN DEAN MOORE

Contents

View Our Climate in a New Light

Care for Your Soul

Cultivate Family Connections

Raise Empowered Kids

Grow Community Connections

Build a Fossil-Free Future

Foreword

Global warming is the deepest, scariest problem that we humans have ever wandered into. We've only known about it in any real way for thirty years, barely a human generation, but it is already changing the biggest physical features of our planet, triggering horrible droughts, fires, and floods and in their wake massive human migrations, and driving many of our fellow creatures toward extinction. One question I get asked often is "Should I even bring a child into this world?" I usually reply yes — and I'll do so with more confidence now that I know I can hand new parents this fine book.

Up to a certain age, we should not tell kids about climate change. It isn't their fault, and it's too big for them to entirely comprehend. Instead, our job is to make our children fall in love with the natural world — and very few people fall in love with the terminally ill. So: hikes, and bikes, and trips to the park, and watching the weather, and keeping animals, and all the stuff that continues to make me happy late into life.

But at a certain point, kids are going to find out that things are not as they should be, at which point it's necessary to give them things to do. Some are things around the house, but what I like best about this book is it doesn't waste a ton of time on light bulbs and the like. Those are obvious, easy — and they won't solve the climate crisis. What will solve it is activism and engagement, and no one is too young to be a part of that. If you grow up going to the occasional march, then you grow up knowing that lots of people are working hard to make the future okay. And that's a deep comfort. By junior-high age, there are kids leading marches and demonstrations — two sisters, seventeen and sixteen, organized the nation of Ethiopia for 350.org in the early days, bringing thousands into the street and changing the high school curriculum.

And what will allow us to survive the effects of climate change we can no longer prevent? Community, I think. For the last fifty years, neighbors have been kind of optional for Americans — if you had sufficient money, you could pretty much dispense with the help of those around you. We built bigger houses farther apart from each other — houses so big, with so many rooms, that kids were in much less contact even with their own families than in previous times. That's why the prescriptions in this book so usefully point us in the direction of connection.

Trust me, this is going to be fun. Humans were actually built for contact — we are social primates, not that far removed from the days when we sat on the savanna picking lice out of each other's fur. It's suburban American life that's the odd anomaly in human experience. So the practices in this book that point us in the direction of gregariousness are among the most important.

There is no guarantee that we'll be able to head off dramatic climate change. But there is a guarantee that the fight against it will build better communities — and better, happier, more connected kids. So read on!

— **Bill McKibben,** author of *The End of Nature*
and cofounder of 350.org

Introduction

Relax — this isn't another light bulb list. It's not another overwhelming pile of parental "to dos" designed to shrink your family's carbon footprint through eco-superheroism.

In fact, drop that light bulb. Stop recycling. Get off your bike, stop gardening, leave your cloth bag in the SUV, and bring hamburgers to the family picnic. Let your faucets leak, leave your windows open with the heat on, and for heaven's sake, don't hang your laundry.

Busy parents — along with everyone else — have been told for years that individual lifestyle changes can stop the climate from spinning out of control, but the truth is they can't. Not by themselves, anyway. This book kicks aside that great myth so we can refocus our efforts on actions that *can* accomplish what those lifestyle actions intend: to build a livable future.

The Climate Crisis Is a Parenting Problem

First, though, let's get clear on one thing: The climate crisis is really a parenting problem. Here's why: We have a problem of too many greenhouse gases in the atmosphere, right? And they're making the Earth heat up dangerously. So, we need to stop adding more gases and reel back in the ones already out there. Reducing greenhouse-gas emissions isn't complicated. It's not easy, but it's *do-able*. In fact, it takes only two steps:

Stop Stupid.

Start Smart.

To stop Stupid, we simply cease burning fossil fuels. Stop deforestation. Stop farming and ranching in ways that unleash carbon. Stop releasing refrigerant gases and methane. Those are the big-ticket items.

To start Smart, we massively expand solar, wind, and wave power, conservation, tree planting, carbon farming, methane capture, local resilience, and more. And we sequester, price, and regulate carbon.

All this requires is a simple switch in mind-set: Turn off Stupid, turn on Smart. Totally do-able. Except that there are madmen in charge who block Smart because they get richer with Stupid.

That's it. It's really that simple. Like drunken teens, adults who should know better don't want their carbon kegger to end.

Most parents, when it comes to their own kids, know how to close down an out-of-control party. They assert authority. They shut off the tap, make everyone clean up and go home, and make new rules to avoid a repeat. We should have done this decisively when we found out in 1988 that burning coal, oil, and gas was cooking our planet. Instead, we let the party continue. We didn't powerfully confront the self-serving denials and lies — the people who said, and still say, "Global warming isn't real, and even if it were, it's certainly not our fault." We've been permissive with Uncle Sam and Exxon for way too long, and things have gotten so far out of hand that we're facing our own — or, worse, our children's — annihilation. And yes, I mean that word. It's really that serious.

That means that, right now, we need to do more than just assert our rightful authority. What's required is full-scale climate revolution.

Revolution may sound extreme, but I'm not suggesting we grab our muskets for an armed insurrection. The climate revolution I'm talking about seeks fundamental political, cultural, and technological changes. And it's already going full steam. It's also fun and rich with potential for personal connection — which is the only kind of revolution that's going to work, frankly. Plus, this revolution is taking place everywhere, all over the world, which makes it easy to join.

What does this climate revolution look like? It's solar farms in Nevada, Danish bicycle highways, and defiant governors upholding the Paris climate agreement in their states.

It's fracking bans, mayors helping fifth-graders pass climate laws, and New Yorkers paid to reduce their electric use.

It's biogas composters in refugee camps, grandparents moving retirement funds from frackers to wind farmers, and Pacific Islanders in canoes blocking coal shipments.

It's college students designing sustainable police stations, climate literacy in public schools, and nuns building chapels in the path of pipelines.

It's South African clerics marching with indigenous leaders, zero-waste apartment complexes, and parents peacefully blocking oil trains running past their homes. It's the Cowboy-Indian Alliance — a group of US ranchers and Native Americans fighting pipelines on their lands together.

The climate revolution is about behaving like mature global citizens. It's that simple. And the solutions, though sometimes difficult, are equally simple: Keep fossil fuels in the ground, drastically cut emissions, and transition quickly to clean, renewable energy. To do this, we can embrace the technology that author Bill McKibben says we need most: "The technology of community — the knowledge about how to cooperate to get things done."

That kind of community invites everyone's participation. You don't need a physics degree, wealth, or connections to celebrities, senators, or CEOs to save the planet. All you need is a beating heart and a desire for authentic solutions.

To be clear, the "climate" I'm talking about refers to our atmosphere, yes, but it includes much more. Each of us inhabits multiple climates — the climate of our individual body and its health; the climate of our home, relationships, neighborhood, school, and job; and our national political and cultural climate. We each have a right to feel safe within all of these climates — a right to a safe workplace, good health care, safe shelter, and freedom from violence and oppression. We have a right to healthy air, soil, and water. And all present and future generations have a right to a livable planet.

All of these climates are deeply interconnected. They're also all badly threatened today — so much so that, in 2015, Pope Francis wrote an encyclical letter addressed to every person on the planet decrying income inequity, war, and degradation of the Earth, "our only home." He spelled out how climate destruction disproportionately punishes the poorest and most vulnerable of the world, and he called for a transition from fossil fuels to clean energy. He then issued a thunderous call for a "bold cultural revolution." Revolution! From a pope!

He gave us a clear moral directive for revolution, but perhaps not being a family man, he was a bit short on the details of exactly how parents might go about waging it. Now, I don't claim to be writing *the* road map, but I hope this can be *a* road map that will help all of us, particularly parents, navigate our way. The worldwide climate revolution boasts no savior or silver-bullet solution. Instead, it's made up of millions of assertions of authority over our own futures that, taken together, are transforming our societies on every level. That kind of transformation requires many maps, and this is the family-friendly one with plentiful games and snacks to sweeten the journey.

Parents Are Key to Saving the Planet

Parents are hardwired to protect their kids. To do so, we sacrifice sleep as we rock colicky babies; we sacrifice money to put food in children's bellies, boots on their feet, and fillings in their teeth; we sacrifice time to help with homework, coach sports, and cheer at recitals. We're downright heroic in countless ways. Any parent I know would donate a lung or take a bullet for even the crankiest teen — and that's what this climate revolution needs: people willing to do what it takes for our kids' long-term well-being. Parents already step up and take responsibility in difficult times. Just as no hero arrives at 3 AM to wash sheets and comfort our child when the stomach flu hits, no one else is showing up to save our kids from climate chaos.

We have to do it. You and I. And the window of opportunity to turn things around is *now.* For lots of reasons that this book will discuss, the time for dithering is over. We need to act now to stop Stupid and start Smart. Everywhere and right away, not tomorrow — and certainly not next year.

And guess what. Parents actually *can* pull off this kind of revolution. That's because we're the bridges to every sector of society. We link the worlds of family and work, of private and public, and of young and old. We parents link our ancestors with our descendants. We're everywhere, and we're raising tomorrow's leaders, who have more at stake than us — they'll live longer and inherit all the problems we've left them. As parents, one of our jobs is to open their minds to this reality in age-appropriate ways, since kids don't always fully understand the peril they're in now. Even if they do, they can't vote or do much about it without our help. Parents can empower kids to start changing and saving their world now.

Furthermore, other adults trust parents. People really do care about families. Grandparents, uncles, coaches, pediatricians, and neighbors share an interest in our children's well-being, one that can transcend political differences and divides. Parents have unique access to people's heartstrings, and it's time to pull them — hard — for our kids' future.

Affecting the kind of real change we need requires people to get together in person. This is the community-building part of the climate revolution, and it's another way families are a perfect fit: Families come together naturally all the time to share birthday cake and stories, carve pumpkins, and play basketball or charades. Who better to gather the global village than families? We're already gathering — and we're *fun*.

Moreover, parents are natural leaders. We're CEOs of small, complex tribes. We manage family finances, calendars, ethics, education, safety, and everyone's mental and physical health. We run all departments in our mission to protect family and home. We simply need to expand our definition of "home" to include the Earth — and Earth's atmosphere.

So, yes, revolution is a lot to ask, but like tending to our feverish child at 3 AM, that's what this moment calls for. And we have a great shot at success. Revolutions are usually successful when people agree with the goals, and survival's a pretty easy goal to rally around.

Why Listen to a Harp-Teaching Frisbee Mom?

I wrote this book because I've spent the last twenty-one years trying to raise healthy kids while also protecting their habitat. When my children were little, my husband, Art, and I rocked green family living — voting, recycling, insulating, fixing leaks, biking, and "living simply." That's what all the experts said would help halt climate change.

But as global warming worsened, I grew increasingly anxious. I kept asking: *How can we parent well in the short-run — keeping our kids happily connected to a good community, education, and experiences — and also in the long run, which means stopping the planet's destruction?* So often, I felt like I had to choose one or the other.

Environmental books, experts, and Earth Day events kept singing the praises of living and buying green, but year after year, the temperature kept rising. Few books proposed structural fixes, such as stopping polluters or reforming an electoral system that lets billionaires buy politicians. Yet, at some point around 2011, it became clear to me (and many others) that climate chaos wasn't some future catastrophe. It was making landfall in our lives. I stopped worrying so much about shrinking my family's carbon footprint and started to focus on shrinking industry's. I wrote articles, displayed political artwork on my front lawn, and cofounded Eugene's chapter of the international climate action group 350.org.

In 2014, when we learned that frackers were trying to export gas to Asia by forcing a pipeline through Oregon, my family, friends, and several 350 Eugene members dramatized the process unfolding in our state with a neighborhood art protest — plastering homes with huge "condemned" signs and running a black, block-long, three-foot-wide pipeline across all of our lawns.

The project united people. Suddenly, I got to know all fifty neighbors living on my block. College kids created hashtags and posted pictures on social media. Middle schoolers hammered pipeline stakes into their own front lawns. Preschoolers drew postcards that people could send our governor in protest of the real pipeline. Because it was large-scale, a little desperate, and done in conjunction with an international environmental law conference in our city, our protest made the front page of the county-wide daily newspaper. My teenage neighbor — who helped hang a twelve-foot-long, bright red "CONDEMNED" sign on her own house — was interviewed on the evening news. The story of our pipeline played statewide on public radio stations for the six days of the installation.

Overnight, hundreds of thousands of Oregonians learned about the proposed export pipeline, and 350 Eugene began to coordinate statewide resistance to the project. But the best part of that pipeline protest for me was seeing how every kid on the block was so thrilled to play a part.

Soon after, three of my son Forrest's friends founded a 350 Eugene/ Earth Guardians climate club at his high school. As a club advisor, I helped them design events and watched them become more empowered as they painted banners, carried them in marches, and gave speeches beneath them. Their families and friends showed up to cheer them on, leaving with gratitude and a renewed sense of optimism. Parents began asking me, "Do you have a list of ten things *I* can do?"

That list grew into this book, which offers ten times that number.

These ideas are offered as a menu of possibilities, not a "to-do" list (which no single human being, much less a busy parent, could accomplish). I hope it will be your invitation to join the climate revolution, the one that's already happening but you might not even know exists. Like the magical world behind Platform 9¾ in the Harry Potter series, the climate revolution is thriving. Yet, despite its tremendous energy and effectiveness, the climate revolution isn't being televised, or even talked about, really, in mainstream media. You have to look for it, but once you do, you'll cross a threshold and see it everywhere. In fact, you'll see the world differently, and you'll recognize opportunities to support this revolution in everything you do, whether you're at the grocery store, at a high school basketball game, or tucking in for family movie night.

One thing I've learned in my own life and through watching other climate-conscious parents is that people won't join the climate revolution if all it means is piling on guilt or adding more to-do items to an already packed schedule; speeding up is neither a fun nor a sustainable way to parent. But if we begin to see our lives through the lens of climate justice, we can revise the things we already do so that our actions more skillfully, efficiently, and *effectively* help create a thriving future for our kids.

That's what I have tried to do, often by channeling my worry into protest and performance art, and my concerns into community building. Another early lesson in this was the Halloween when I decorated our front lawn with our usual ghosts and spiderwebs, but then added "The Graveyard We Need." I made headstones that named many of the things that were tormenting me, including "Mountaintop Removal," "Climate Denial," "Tar Sands," "Oil Addiction," and "Our Fear and Silence." As I hammered in the stakes to prop up those headstones, I felt nervous about how this statement would be received by our neighbors, but I also felt more alive. Then, on Halloween night, as two hundred trick-or-treaters and their families trooped to our front door, I was pleasantly surprised by how many complimented and appreciated our display.

What I learned then — and what I have continued to learn during the various art installations, protests, and lectures I've led since — is that most people welcome the conversation. Most people are delighted — or at least intrigued — by a disruption in business-as-usual, especially when it's playful and nonthreatening. Most people, in some way, embrace the idea that "home" includes the rivers we drink from, the soil that feeds us, and the climate that shields us.

Most importantly, what I have learned is that a sustainable ecosystem and climate requires sustainable parenting. To me, sustainable parenting means that we don't just pour our loving attention into our own children for eighteen years, preparing them for success in college and the job market. It means putting on the climate lens for every interaction so that we can prepare them to thrive in the natural world that sustains their lives — and in the democracy that shapes the quality of those lives. Sustainable parenting means taking the well-being of *all* children into account — not just our own — when deciding how to live in the world and what to model around the ethical use of the world's resources. Finally, to me, sustainable parenting means loving our children for their whole lives. That requires acting in ways that let them feel our love now and that ensure they'll feel it after we're gone, when they are thriving in the world we've pulled back from the precipice.

Use This Book (Don't Just Read It)

Busy parents hardly have time to read, so I wrote this book in a way busy parents can use. Of course, you can read this start to finish, and I organized the chapters to unfold in what I hope is a logical and helpful way. Or you might adopt the fortune-cookie method — keep the book nearby and, whenever you have a couple of minutes, crack it open and see what you get. Most chapters are short enough to fit into the average bathroom break parents with young children allow themselves. The goal is to fit climate revolution into busy family life in ways that help us stay balanced and that leave ample time for good sleep every night. A few chapters are longer than others, but all are written to both inform and entertain — and we can certainly use more levity in climate conversations, can't we?

In addition, and above all, each chapter offers practical advice for parents with limited time and resources. Even if you have little money and only an occasional five minutes to spare, you will find effective and satisfying ways to support the climate revolution while also empowering kids. The book also includes lots of advice for those with more money, time, and heroic enthusiasm than they realized.

I included various activities for all ages, temperaments, and abilities. There are plentiful actions appropriate for introverts, extroverts, homebodies, children with special needs, parents, grandparents, teachers, counselors, doctors, single people, and more, and readers will find ideas for every lifestyle, whether they live in an apartment, farmhouse, or suburban home. The climate revolution needs everyone, everywhere.

Some chapters do present ways we can minimize our individual and family climate impacts. Despite my earlier "drop-that-light-bulb" remark, I do suggest replacing light bulbs and doing all the little things to shrink our personal carbon footprints. However, to me, the main reason to do the "little things" is that they are part of being a climate-conscious parent: They model for our kids that we care about the Earth and other living beings. Also, more than we sometimes realize or are aware, everyday actions — such as biking to school or buying palm-oil-free pet food

— can be just as easy to do as to not do. Kids need to get to school and we have to feed Fido, anyway, so if time and money allow, we may as well make Earth-friendly choices.

That said, I wrote this book from the deep conviction that, if all we have on any given day is three minutes to aid the climate revolution, and our choice is either to scrub out our empty peanut butter jars (for recycling) or to call our senator about a key climate decision, it's more effective — and more important at this point in the climate crisis — to make the call and throw the dirty jar in the trash. If you have to make a choice, always choose system change over light-bulb change, and if you've got energy left after taking a revolutionary action or two, go right ahead and change that light bulb.

Furthermore, I hope you will use this book to connect with others and to be inspired and energized by all the people, especially the parents and kids, who have already eagerly leaped into the climate revolution. As I did my research, I was astonished by the tens of thousands of dedicated people working full-time — and literally millions more working part-time — on some aspect of climate recovery. So many people worldwide are giving their hearts and souls to ingenious inventions, cutting-edge building design, investigative journalism, environmental entrepreneurship, pioneering litigation, pro-planet artwork, good old-fashioned rabble-rousing, and every manner of research. Countless people are initiating compost programs, challenging oil companies in court, testing solar roadways, pushing for wind farms, building better energy-storage systems, teaching environmental democracy, closing coal plants in their cities, experimenting with soils and seeds, and sharing indigenous wisdom. Meeting and learning about these people heartened me beyond measure. Though our climate situation is dire, I don't lie awake worrying over my children's future as I did a decade ago — partly because I'm more actively engaged in solutions, but mostly because I don't feel alone anymore. I hope their stories and offerings do the same for you.

I invite you to use this book to inspire and empower your own kids. If parents do nothing else but that, then they are putting a firm shoulder to the climate revolution. And isn't that what all parents want, anyway — empowered children? We want children unafraid to learn, question, make connections, help others, speak out, lead, and take responsibility.

Children brave enough and supported enough to fight for their vision of a better, fossil-free future: one defined by electric buses — not diesel fumes and cities choked with smog. One without oil trains snaking through communities like ticking bombs. One without fracking toxins that poison our drinking water. One without oil spills on beaches or fish too polluted to eat.

None of these things are necessary because burning fossil fuels isn't necessary, not with the clean and renewable energy solutions that we already possess, which are only becoming cheaper, safer, and more effective. The main thing standing in our way is uncertainty about our own power. We just have to give ourselves the authority to flip the switch from Stupid to Smart.

Finally, I hope by using this book, you find that waging climate revolution improves all the climates that define our lives. This particular revolution has a delightful cascade of cobenefits. For one, it's healthier. Reducing car trips and biking more, encouraging active play, and eating plant-based diets are better for our health. Reducing materialism is good for family budgets, closets, and souls. Green jobs are healthier for workers and better for local economies and resilience. Active engagement in climate solutions connects us to one another and improves our communities and society. And doing it with children energizes the conversation. I've seen it time and again: Empowered, passionate children inspire others — and make the world a better place.

As it turns out, pretty much everything that's good for the planet is also good for children, which is why philosopher Kathleen Dean Moore writes that "although environmental emergencies call on us to change, they don't call on us to give up what we value most. They encourage us to exercise our moral imagination and to invent new ways of living that lift the human spirit and help biological and cultural communities thrive."

This is my invitation to you. Ours is a time of uncertainty and anxiety over the future, but it's also an extraordinary time to be alive and raising the next generation. We're seeing unprecedented international cooperation and acts of great imagination, as well as opportunities to join in the epic global transformation. I'm not pretending there's any guarantee of success. I am saying, though, that even if we don't know how things will turn out climate-wise, we can gather our children — the sooner the better

— and join countless other families actively and lovingly doing what they can to build a fossil-free future.

In our homes and in the world, we can stop Stupid and start Smart in countless ways, from the bottom up, until we transform not only our own attitudes, lifestyles, and communities but the larger political, economic, and cultural forces shaping our kids' futures.

Let's give it a try. Let's show the world — and especially our children — how fun, unifying, and empowering it can be to save the world.

HARMONIZE
FAMILY LIFE-WAYS
WITH EARTH-WAYS

1. Get Clear on Why There's Hope

A lot of my colleagues have now said it's too late. We've passed too many tipping points to go back. My answer is thank you for the message of urgency.
— DAVID SUZUKI

I'll never forget the moment my optimism returned. It was 2013, yet another year of record-crushing heat waves, droughts, and forest infernos. The concentration of atmospheric carbon had sailed past the safe limit — 350 parts per million (ppm) — to an unprecedented 400 ppm. Alarmed scientists begged leaders to cut carbon pollution. In response, most leaders chanted, "Drill, baby, drill!"

We citizens kept losing campaigns to protect ecosystems. How could we possibly slow, much less stop, the death march for polar bears, honeybees, apple trees, kittens, and even our own children? I lived with a near-constant lump in my throat.

Then my daughter invited me to a film about her friend, Kelsey Cascadia Rose Juliana, who was suing the government for violating her right to a livable planet (for more on this continuing story, see "Sue the Grown-Ups!," page 244). The short documentary, produced by Our Children's Trust, ended with a "prescription" for balancing the climate by 2100:

1. Cut global emissions by 6 percent yearly, starting in 2013 (the target is now 9.9 percent; it rises until we start making those cuts).
2. Begin massive reforestation.
3. Improve farming practices to sequester carbon.

Lights came up. People clapped. I scribbled the prescription, wondering, *Is this really a fix-it plan to avoid wholesale catastrophe?* At home, I researched and found that, indeed, an international team of eighteen climate scientists, led by Dr. James Hansen, then director of NASA's Goddard Space Institute, had determined that if we accomplish those three things, climate stability *is* achievable. Not cheap or easy — but possible. We *can* get the nasty gases down again to the safe level of 350 ppm.

I hadn't understood that. I thought that when scientists raised their usually calm voices and said, "Holy crap! We're at 400 ppm!" they meant we'd *always* be there or higher. That we couldn't subtract carbon.

It turns out we can. We can draw carbon down from Earth's atmosphere and lock it into soil and trees — by planting new trees, by refraining from cutting more down, and by using many other methods you'll read about in this book. If we do that while also drastically cutting emissions, we can restore climate stability by the end of this century. The climate will keep heating up — due to humanity's inaction over the last thirty years — but how much and for how long depends on what we do next.

Whether the climate heats by 1.5 degree Celsius — or by three or four times that — depends on the decisions that governments, corporations, societies, and individuals make right now. Do those three things, and every year we can make things a little better. Don't do them or wait too long, and those same scientists say humanity will lose its opportunity for climate stability.

The good news is that the prescription — with its clear targets and the reassurance that we can still turn things around — has fired people up worldwide to demand science-based climate action. Even countless religious leaders now say we must keep most fossil fuels in the ground.

That gives me tremendous optimism — and compels me to help. I invite you to jump in with your family — in ways that work for you — and have a blast.

That's what this book is about.

If you have...

Five minutes: Tuck this book next to your bed, in your car, or in the bathroom. Read one chapter at a time.

Thirty minutes: Visit the Our Children's Trust website (http://our childrenstrust.org). For a good overview of how scientists' "prescription" is being used by children to fight for a livable climate, watch, with older kids, the short film A *Climate of TRUST*. Look for it under "Short films."

Six hours: Binge-read this book. Think about your skills and passions — and where you might want to plug in to the grand adventure.

2. Herd Your Family Together

You could argue that central heating played a part in the start of the disintegration of the family.
— ELEANOR JOHN, head of collections at
London's Geffrye Museum of the Home

One way we deepen family connection is by keeping bedrooms unheated, gently forcing a cozy nightly living room scene. Aside from saving money, shrinking our carbon footprint, and sleeping better in cool rooms, the heat-only-the-living-room trick also allows for easy monitoring of screen use and homework progress.

Mostly, though, it's congenial. When both our children lived at home, we happily stepped over our daughter, Zannie, parked with art supplies or Spanish homework in front of our fireplace, since it was nice to interact with her instead of her closed door. She'd randomly ask, "Does *mariposa* have one *r* or two?" and chat frequently with her younger brother, Forrest, as he built four-story card houses.

The key to this idyllic incubator for young scholars, we've found, is for one parent to be interruptible. If the other needs focus, that parent can retreat to a bedroom or my studio. Usually, though, everyone prefers to stay with the herd.

We regularly hear parents helplessly bemoan the kid they don't know anymore, who's cloistered upstairs doing god-knows-what on-screen. This lost-kid scenario has three main culprits, the first being central heating. When that pours into the second culprit — private bedrooms for every family member — children can do what few could before the 1980s supersized our homes: detach from family life.

Completing the trifecta is the electronic device taken into the cozy kid cave. At best, unmonitored screens in bedrooms stoke consumerism and encourage kids to inhabit separate worlds. At worst, they leave children vulnerable to disturbing websites and online pedophiles, con artists, cyber-bullying, pornographers, and relentless and predatory marketing, as well as all the ills associated with excess screen time (see "Unplug the Kids," page 145).

If you want to...

Save a little money (and the climate): Turn off the heat and/or air-conditioning in bedrooms.

Not buy desks for the kids' rooms: Use the kitchen table or couch, or create a living room study-hall atmosphere that prioritizes scholarship — not TV. Let kids store a reasonable stash of books and study supplies in a designated spot in the living room (or wherever you have space near the study area).

Save a lot of money (and avoid screen strife): Have only one computer available in the evening, if possible, for the whole family, at a work station. When a child reaches high school, give them first dibs on it for nightly homework. In our house, this pecking order helped make it clear that our high schooler's work trumped any other use of the computer, including my email and anyone's recreational time. When Art and I could afford separate laptops, the family laptop was suddenly freed up more often, creating new battles over its use on school nights. If I had a do-over — and knew then that easy-access YouTube was headed to our living room couch — I'd save my money and go back to sharing one or two computers for a family of four at night at a work station. The inconvenience of sometimes having to wait in line for a computer would have been preferable to the much harder parental job of limiting screen time for its own sake.

Save family harmony: Together, analyze how you use your space and what connects or separates your family. Kids are usually delighted to be consulted on decisions traditionally left to grown-ups, and they often hatch very do-able plans for smarter use of space. Now that my kids are young adults, we negotiate everyone's use of phones so that people can get their work done and plans made without being immersed in phones, especially when others want or need to connect.

3. Plant Trees!

Stop talking. Start planting.
— Plant-for-the-Planet's motto

One of my favorite photos of my daughter as a toddler is of her eating lunch at the base of a giant redwood tree. She is blissfully happy. We all were that day, maybe because being in the presence of thousand-year-old trees in a sun-dappled, sweet-smelling grove just brings out the best in people.

And those magnificent giants don't just offer beauty, shade, protection from erosion, and wildlife habitat. They capture a staggering amount of the carbon humanity has flung skyward. It's miraculous, really, the way trees reach into the heavens and gather carbon in their leafy arms. Which is partly why we need lots more of them. According to the book *Drawdown*, edited by Paul Hawken, the combined impact of protecting forests from logging and restoring forests worldwide makes up "the most powerful solution available to address global warming."

Wow. Okay! How do families help with that? We can raise little tree-huggers. Protect forests. Plant trees. Lots of them.

That may be one of the simplest and most enjoyable actions families can take. Just acquire and plant trees. Even better, grow a tiny forest. This is a new, successful method catching on worldwide: Start a dense, multi-storied native forest in an area as small as six parking spaces, and after two to three years, it becomes maintenance-free — and because it's thick, it sequesters a surprising amount of carbon.

As a family, join or support the group Plant-for-the-Planet, a kid-led nonprofit organization whose goal is to plant one trillion trees by 2020. The group and its sixty-three thousand kid "ambassadors" have already planted fifteen billion. When the group launched in 2007, nine-year-old Felix Finkbeiner described their simple strategy:

"Children could plant one million trees in every country on earth and thereby offset CO_2 emissions all on their own, while adults are still talking about doing it."

On the policy front, families can demand that our government protect North America's forests and reduce logging. We can also advocate protection of virgin tropical forests, such as in the Amazon basin, which are key to keeping carbon locked up in our greenery. International agreements such as the Paris climate agreement can protect them by paying forest communities not to cut them.

We can also boycott tropical hardwoods. Logging drives tropical rainforest destruction, and the United States is the biggest importer of tropical hardwoods. These include mahogany, Caribbean pine, ipê, rosewood, teak, and ramin. In our homes, these woods are used in furniture, veneers, cabinetry, instruments, decking, fence posts, tool and cutlery handles, picture frames, pool cues, moldings, flooring, beams, pencils, and baby cribs.

But today? Get outside, have a picnic under a big old tree. And don't forget to take a picture of your little sprout. She'll be mature before you know it.

If you want to...

Inspire kids: With young kids, read *The Lorax* by Dr. Seuss and *Wangari's Trees of Peace: A True Story from Africa* by Jeanette Winter.

Visit Plant-for-the-Planet (www.plant-for-the-planet.org), let kids become ambassadors, and watch the video in which founder Felix Finkbeiner movingly challenges adults at the United Nations (www.youtube.com/watch?v=tGLtkbaeupI).

With teens, watch *Pickaxe — The Cascadia Free State Story* (http://freecascadia.org/pickaxe-the-cascadia-free-state-story): This documents the eleven-month battle led by the group Cascadia Earth First! to save an old-growth forest in Oregon.

Visit Drawdown's gallery of climate solutions (www.drawdown.org/solutions-summary-by-rank), and discover all the ways trees are key.

Plant a tree: Plant a tree in your backyard. Some families decorate potted trees for Christmas, then plant them. Or plant lots of trees in a vacant lot or unused land. Visit Afforestt (http://afforestt.com) for information and to watch Shubhendu Sharma's Ted Talk "How to Grow a Forest in Your Backyard."

Save the forest: Join and support the Rainforest Action Network (www.ran.org), which "preserves forests, protects the climate, and upholds human rights." (Not to be confused with Rainforest Alliance, a certification organization that has been criticized for corporate greenwashing.)

Buy wood sustainably: Avoid tropical hardwoods and buy recycled wood or sustainably sourced wood products certified by the Forest Stewardship Council. Consult the Natural Resources Defense Council's guide "How to Buy Good Wood" (www.nrdc.org/stories/how-buy-good-wood).

4. Lift Moods and Grades

The health benefits of cycling are twenty times greater than any risk involved.
— MIKAEL COLVILLE-ANDERSEN, Danish urban mobility expert

My kids' high school is a mile from home. Barring a blizzard, broken femur, or Halloween costume wider than the sidewalk, I expect my kids to use their legs to commute.

This isn't primarily to save the planet, though that's a benefit. It's because exercise vastly improves moods, grades, classroom behavior, and health. Having your child walk to school saves money, time, the planet, and our atmosphere (less pollution), while *also* improving your child's mental and physical health.

Why, then, do more of us than ever drive kids to school? Fifty percent of parents drive from a half or even a quarter mile — *three city blocks* — from home.

Two reasons: We want our kids to be happy, and perhaps more pervasively, we're afraid.

I'm certainly not immune to fear. When my kids first trekked alone to school, I feared kidnappers, and I dreaded the flu when they refused to wear raincoats. I still worry they'll be killed by texting drivers, but I've learned to manage my fears, and you can, too.

Of course, when distances are far and there's no bus, train, or safe biking route, kids may need to be driven. But otherwise, try to conquer those fears one by one.

If you fear...

Strangers: Parental fear about kidnapping is nearly universal. However, for kids under fourteen, the risk of death by stranger abduction is far lower than that of death from flu, heart disease, plane crash, lightning, or health crises associated with a couch-potato lifestyle.

Solutions: Try a "walking school bus" or "bike train," in which one adult accompanies a group of children on foot or bike. Safe Routes to School (http://guide.saferoutesinfo.org) offers advice for how. Lenore

Skenazy, author of *Free-Range Kids*, offers more advice on her website (http://www.freerangekids.com).

Pneumonia: Kids are notorious for underdressing in bad weather, be it snow, rain, ice, or hail. Is it okay to let kids brave the elements?

Solutions: Let them be wet and cold. They'll survive. Most colds and pneumonia are caused by viruses — not by being cold. Kids are only at risk if they get hypothermia — unlikely while walking to school. If your kids refuse a coat, at least insist on a healthy breakfast before they get chilled.

Respect, don't ridicule, the urge to look cool. Let kids choose their own rain gear, umbrellas, and so on. Until they stop growing, buy sale and secondhand gear, which is financially easier to replace.

Death by car: Everyone risks injury from cars.

Solutions: Teach walkers to stay on sidewalks and how to safely cross streets.

Teach kid cyclists to obey traffic rules and to ride alertly — and defensively — around cars. Ride with kids to find the safest route to school. Sometimes, a busy artery with wide bike paths and heavy bike traffic is safer than narrow, quiet side streets.

Use well-fitting helmets and well-maintained bikes, with fenders, a horn, a light, and multiple reflectors (on backpacks and coats, too).

Back problems: Heavy backpacks are an issue whether your child walks or bikes to school.

Solutions: Keep textbooks at home. Ask teachers to loan copies in class, or buy a used copy (online) to keep at home.

For cyclists, carry books in a bike basket or pannier, or attach a milk crate (see Pinterest for inspiration).

Replace heavy binders with light folders and spiral notebooks, and regularly purge backpacks of unnecessary papers.

Social exile: Once, my normally easygoing seventeen-year-old daughter, Zannie, refused to wear a helmet (see "Help Kids Feel Comfy," page 95). She explained, "I'm okay being a nerd, but I won't be the *only* one biking to school in a helmet." A power struggle in this situation was clearly doomed.

Solutions: I let her choose a new cute helmet and whether to wear it or not. (I'm happy to report that she does.)

5. Cultivate a "Used-Is-Cool" Culture

Fast fashion is fast food. Empty calories that make us feel full. Factories full of mistreated workers. Rivers full of toxic chemicals. Closets full of disposable wears. Landfills full of yesterday's garments.
— ZADY CLOTHING COMPANY

One week before Forrest hit middle school, he announced, "I'm *done* with thrift stores." He wanted brand name and brand-new. For the previous decade, he'd happily worn whatever castoffs I collected from neighbors, cousins, and garage sales. But when I piled his too-short pants into the giveaway box and made our back-to-school shopping list, he made it clear he would not trot around to stinky thrift stores and paw through stained clothing worn by *other people*!

I took him to the mall and gave him a seventy-dollar budget. He tried on fifteen pairs of jeans, experiencing firsthand how buying new requires almost as much trying-on effort as buying used does, and he chose one forty-dollar pair and two fifteen-dollar shirts. That was it, no more school shopping. It was outrageous, of course, since he'd outgrow those expensive jeans within a year, but he got his brand-new, brand-name clothes.

Over the next months, he didn't reach for those new clothes much, and he commented on it. Once the allure faded, they became no more special than any of his other clothes because they *weren't* special. Just expensive. And now used.

The next August, I offered to buy both kids as much clothing as they wanted — provided it was used. That, along with Zannie's enthusiasm, got Forrest through the thrift-shop door. Two hours later, he emerged with two bags overflowing with everything he wanted, including a couple of oddball jerseys and hats for experiments in his evolving personal style. To ensure he had happy memories of thrifting, I topped it off with fish and chips and root beer from a nearby food cart, where both kids reveled in their great "scores" and swore off the mall forever.

My used-clothing campaign isn't just to save money or reduce our ecological impact, though it does those things. We also want to keep our children's materialistic expectations low. We're in the shrinking middle class — whose real wages haven't risen since 1978, despite skyrocketing health care, food, housing, and education costs — and I want our kids to know how to thrive on the cheap.

So we model a love of secondhand stuff. When the kids were little, we went to countless garage sales, keeping their books, games, toys, and sports equipment fresh and age-appropriate. We buy used furniture, beds, bikes, homes, dishes, cars, instruments, electronics, and even fabrics and art supplies.

We do buy many things new — underwear, socks, toiletries, small appliances, linens, kitchen utensils, and the occasional laptop, raincoat, or tent — but we do our best to buy sustainably to minimize the various global nightmares of overconsumption (pollution, sweatshops). Sometimes we fail.

But I have to say, emphasizing secondhand has accomplished what we hoped it would: Our kids aren't driven to consume, especially new clothing. In fact, last week, my increasingly nonconformist son showed me his new (to him) T-shirt stenciled with "Try Everything" and announced: "From now on, I'm only buying these weird, one-of-a-kind handmade shirts from thrift stores."

Fine by me.

If you have ...

Young children: Help them imprint happily on garage sales. Bring muffins and juice. Bike or drive by slowly, only unloading for promising sales. Go early, visit one to three sales, and call it good. If you're on a mission to buy specific items, go without children.

Middle schoolers: Lavish attention and personal shopper service while they try on thrift-store outfits.

Splurge at an art supply resale store, found in most cities. Throw an art party.

With them, visit Zady's well-researched, interactive exposé of the fashion industry's human and ecological costs (https://zady.com/the newstandard/our-mission).

Teens: Teach them to safely purchase items on Craigslist. See "How to Buy and Sell Safely on Craigslist" on Lifewire (https://www.lifewire.com /how-to-buy-and-sell-safely-on-craigslist-2487155).

Used clothes: Don't landfill — donate! Search Earth911 (http://earth911 .com) to find clothing recyclers near you.

6. Give Up One Thing

My ten-year-old knows that, rain or shine, she's riding her bike to school for the polar bears.
— MY FRIEND MELISSA HART

We live like royalty, my family recently decided. Machines water our gardens, guard our safety, feed our animals, cook our meals, flush our waste, heat our bathwater, carry us to market, and warm, cool, entertain, and, good heavens, even massage us.

We've got ancient kings and queens beat, and we're not even compelled, as they were, to feed, house, and intimidate scullery maids, litter bearers, and sword-wielding guards. Until recently, what monarch could take to the sky, as any commoner can do today, and arrive on a distant continent after an afternoon spent watching movies and eating pretzels on a nice soft seat?

We live lives of extraordinary enchantment. Yet, we complain. We want more. We want faster, cheaper, better, and we want it *now*. That kind of hyper-luxury requires a cruel sacrifice, though. We sacrifice the environment, the living conditions of millions in poor nations, and the future well-being of our own children.

Unless, that is, we don't. What if, instead, we decided to sacrifice just a few of our royal riches? We can start by being more mindful and ask, "How can I meet my needs without destroying the natural world?" Perhaps we could shift to a smaller home, buy less stuff, or eat diets based mostly on sustainably grown plants.

We can recognize the unique urgency of this moment and offer a personal sacrifice, what author Mary C. Wood calls a "down payment" toward recovery efforts. She advocates renouncing air travel for a year or two during this critical window, but we could broaden the concept. Just as Catholics give up something for the forty days of Lent each year — such as bad habits or cherished treats — we can mindfully choose one thing we do or use that we know harms the climate.

If the word *sacrifice* feels severe, we might reframe the concept as radical generosity, or even as an honor, the way members of the military view their service. However we frame it, let's include our children, who might enjoy the grown-up feeling of subtracting one thing on behalf of something larger than themselves. Renouncing something as a family can be bonding during moments when that sacrifice is challenging for one member, and fun when the allotted time ends, and we get to enjoy our riches with a new sense of appreciation.

If you want to give up...

Apathy: Read the hilarious *Grist* online magazine (https://grist.org). Start with the "21-Day Apathy Detox" and sign up for "Briefly," their pithy daily summary of climate news.

Flying: To connect with far-off family, try Skype, letters, FaceTime, phone time, and care packages.

Environmentally expensive food: Give up red meat. Avoid coffee, sugar, peanuts, and other tropical luxuries that are planted annually (and therefore have a higher carbon footprint than perennial tree fruits).

A new car: Make do with the clunker. Or sell both clunkers and buy one electric car — or an electric bike.

A vacation: My friend Laurie canceled her jet-setting vacation and donated the money to Oregon's pipeline battle.

Home expenses: Delay that remodel — at least for a year. Then, remodel using salvaged materials as much as possible.

Silence: Speak out about the massive climate impact of our endless wars and bloated military.

Bystanding: Consider cutting back your work hours, living on savings or more frugally, and volunteering to help lead the climate movement during the critical next year or two.

Money: Donate to victims of climate-caused disasters. Do a little sleuthing first. (The American Red Cross has been criticized for its conduct and spending choices after several hurricanes.) Local churches or nonprofits may be more effective at directly assisting victims.

Assumptions: Read Barbara Kingsolver's *Animal, Vegetable, Miracle* about her family's year of forsaking foods from more than 250 miles away (bye-bye chocolate and sugar!). Or watch *No Impact Man* by Colin Beavan, whose family tries to eliminate their carbon footprint.

7. Ditch the Diaper

Like all mammals, human babies instinctually resist soiling themselves, their sleep space, and their caregivers, and they clearly communicate about it from birth.
— ANDREA OLSON

My friend Bethany held her newborn over the potty, waiting for the flow of pee.

You must be kidding, I thought. *Juniper's seven days old!*

Two seconds later, pee hit the bowl.

I called it lucky coincidence until it happened again twenty minutes later, and on every subsequent visit. Seventeen months later, Juniper was diaper- and accident-free — in less than half the time it takes most children in the United States.

Welcome to the world of "elimination communication." I had no idea diaper freedom was possible when I had babies. Eager to avoid polluting, we slogged away with diaper services and with cloth diapers we washed ourselves. When rash tormented our cloth-wrapped babies, we turned to the dreaded disposables.

I wondered what diaper-free indigenous cultures knew that I didn't, until Bethany told me, "You pay attention to your baby's signals. They're telling us, with their sounds and movements, when they need to poop and pee. Then you hold them over the potty. It's like house-training a puppy, actually."

Planet-wise, reducing diaper use or going diaper-free is a no-brainer. We save the water and detergent used to wash cloth diapers and reduce the number of diapers that are landfilled — 24.7 billion annually in the United States alone. The average petroleum-based disposable takes between fifteen and five hundred years to disintegrate — depending on conditions and who you ask. Toss in dyes, perfumes, volatile organic compounds, and dioxins — known carcinogens — and you've got a toxic cocktail outgassing in our landfills. And on our babies 24/7. One 2010 study stated, "Disposable diapers should be considered as one of the factors that might cause or exacerbate asthmatic conditions."

Moreover, in disposables, the material used to absorb all that pee is primarily wood pulp mixed with super-absorbent polymers. Just when kids need us to plant trees, we're instead pulping hundreds of thousands yearly and shoving them between babies' legs.

And *that*, a 2012 study found, makes babies fall over three times as often when they're learning to walk. Diapers introduce bulk between the legs, forcing the legs apart and making it harder for babies to balance. Walking diaper-free, researchers concluded, is superior to both cloth and disposable diapers.

That convinced Bethany, a healthy-movement therapist, to learn elimination communication (EC). She explains, "Reduced diaper use sets babies up right from the start for a more fluid gait and healthier lifelong alignment."

Before the advent of disposables in 1961, 92 percent of kids were potty-trained at eighteen months. In 1999, that number had dropped to *4 percent* for two-year-old children.

Then there's money. Three years of conventional diapering costs roughly $3,000. EC requires only a little gear upfront — potty, cloth wipes, undies, and diaper belts — for about $100 to $200.

What are EC's downsides?

It takes time to learn (like breast-feeding), so it's a bit messy.

Babies and toddlers sometimes backslide (EC is easier with wipe-able floors than with carpet).

Parents are interrupted more, since tiny bladders need frequent emptying.

Babies must wear little pull-down undies and pants, so no cute overalls or onesies!

The cool thing, Bethany says, is that you can use EC when you want. "It's not all or nothing. Doing it even sometimes is beneficial."

If you want to...

Reduce diaper use or go diaper-free: Join the international support network at DiaperFreeBaby (http://diaperfreebaby.org). Visit Go Diaper Free (https://godiaperfree.com), which is hosted by mother-of-four Andrea Olson. Olson also runs Tiny Undies (https://tinyundies.com), which sells cotton underwear for EC babies. And get an "Eco-Friendly BecoPotty," a *biodegradable* training potty (sold on Amazon).

Switch to cloth: Visit All About Cloth Diapers (http://allaboutcloth diapers.com).

Use Earth-friendly disposables: Tricky to impossible, depending on your definition, but certain disposables have fewer bleaches, dyes, and toxins than the mainstream ones. Read in-depth reviews on Robin Hill Gardens's blog (https://blog.bolandbol.com/product-reviews/diapers-review).

Try a compostable diaper service: Search for this service in your city. Properly composted at commercial facilities, soiled diapers break down 100 percent in scant weeks into nutrient-rich topsoil used in landscaping.

8. Give Palm Oil the Back of Your Hand

Are your cookies causing orangutan extinction?
— RAINFOREST ACTION NETWORK

My mother never bought Oreos, but I forgave her after a wonderful discovery. Standing on the kitchen counter, I could remove the tub of Crisco oil hidden above the fridge, mash the white grease with sugar, and smear it on chocolate wafers I found hidden in the pantry. I didn't know then that my faux-reo had a dark side.

Crisco's culprit — also in processed foods, cosmetics, cleaners, pet foods, deep fryers, and biodiesel fuels — is palm oil. It's the real Earth-killing deal, what *Scientific American* calls "the other oil problem." On hell's baseball team, palm oil would have all bases covered: On first, razed tropical forests. On second, countless endangered species (hardest hit are orangutans, who lose their habitat to deforestation). On third, destruction of indigenous communities, complete with child labor, forced labor, sex trafficking, and all the usual corruption and abuse that occurs when industry rapidly expands in countries with weak environmental and human-rights laws.

What really knocks it out of the park is that palm oil is a climate terrorist. Multinational companies slash carbon-rich tropical forests for the valuable timber, unleashing massive amounts of the greenhouse gas, then torch the lands underneath, often peat swamps even richer in carbon than the forest. The fires — visible from outer space — can smolder for months. Indonesia's pristine rainforest is rapidly being decimated by logging and palm oil production, killing thousands of orangutans yearly; due largely to forest fires, Indonesia is the world's third-largest greenhouse-gas producer.

What can Oreo-loving families do? The Roundtable on Sustainable Palm Oil (RSPO) attempts to certify growers using sustainable methods, but its standards don't ban destruction of forest or peatlands. Critics accuse RSPO of greenwashing an unsustainable industry and recommend boycotting palm oil altogether to decrease demand; others argue palm oil's here to stay, especially as consumers flee partially hydrogenated oils, or trans fats, so we should use the least-offensive production methods.

If you're animal-lovers, you might want to skip or at least reduce palm oil use. We're trying that route. Recently, as Forrest studied physics, I announced, "I won't buy Newman-O's or Trader Joe's bars anymore because they kill orangutans and the planet."

He glanced up, said, "Fine," and returned to the glorious world of antimatter. It's not always that easy — palm oil is in countless products, especially snacks. But given the choice, kids usually pick the planet over junk food. Back in my Crisco-eating days, I'd have surrendered my faux-reos to protect my distant ape cousins, especially if given an alternative.

We snack on more fruits, toast, nuts, popcorn, and home-baked goodies now, and when we can't, we seek palm oil–free treats. That means avoiding many processed foods, especially creamy fillings and chocolate coatings. Skipping processed foods is a good idea, anyway, budget- and health-wise. For cooking, we use organic olive oil, butter, Fair Trade Certified coconut oil, and vegetable oils. Lard's also an option. We deploy the 80 percent rule: The 20 percent of the time we travel, socialize, or host, we go with the flow. The rest of the time, it's bye-bye palm oil.

If you have...

Five minutes: Challenge kids to a Pantry Palm Oil Hunt. Who can read labels and find the most products with palm oil?

Learn palm oil's many aliases and those of its derivatives, including: palmate, PKO or palm kernel oil, stearic acid, sodium lauryl sulfate, cetyl or octyl palmitate, elaeis guineensis, and names ending with palmitate.

Watch the slick (fake) Doritos commercial "A Cheesy Love Story" that aired during the 2016 Super Bowl, linking tropical deforestation with the chip (www.youtube.com/watch?v=VPlxNhEc2lA).

Switch to chips cooked in a different oil.

A little more time: Join Rainforest Action Network's campaign targeting "Conflict Palm Oil" (www.theproblemwithpalmoil.org).

Find palm oil–free products. Cheyenne Mountain Zoo (www.cmzoo .org) has an online shopping guide, shopping app, family resources, and links for political advocacy.

Thirty cents a day: Adopt an orangutan through Orangutan Outreach (https://redapes.org).

Money to buy some rainforest: Buy some rainforest! Contact Orangutan Foundation International (https://orangutan.org).

9. Bag the Cow

If cattle were their own nation, they would be the world's third-largest emitter of greenhouse gases.
— PAUL HAWKEN, *DRAWDOWN*

Before I discuss how to reduce the huge climate hoofprint of animal agriculture, I need to say this:

I regret being vegan during my first round of pregnancy and nursing.

It's not that my firstborn isn't healthy — my eight-pound bundle of joy today stands at five-foot-nine and summits mountains with the best of them — but *I* suffered. No matter how much flax oil, vitamin B, seaweed, nettle tea, and steamed kale I ingested, I was chronically anemic and finally resorted to prescription iron supplements.

I wish I'd listened to friends who said, "Sweetheart, you're building a person. Have a burger." I'd have improved my health, mood, relationships, and parenting. After three years of ass-dragging with a toddler, I ditched my eighteen-year vegetarianism, quickly regained energy, and safely birthed and nursed another baby.

I wish women didn't pressure themselves, as I did, to help the environment and reduce animal suffering through denying ourselves — or our children — proper protein, iron, B_{12}, and essential amino acids during those fleeting, crucial baby-building years. That said, most people thrive on a mostly plant-based diet. My family is now "reducetarian," which means we try to eat fewer animal products.

But that's not my main point. The most important thing parents can do to cut emissions isn't to go vegan or reducetarian, though that definitely helps. It's to reform animal agriculture.

A 2009 World Watch Institute study estimates that "livestock and their byproducts actually account for *at least*...51 percent of annual worldwide GHG [greenhouse gas] emissions." *Fifty-one percent!* And that number's rising with our global population.

It doesn't have to be that high. We already know ways that industry can stop spewing so much carbon, nitrous oxide, and methane into our ailing atmosphere. We just need lawmakers to legislate them. Leaders can do the following:

1. Stop subsidizing animal agriculture. Taxpayers pay roughly $38.4 billion yearly to help producers of animal products. When we factor in externalized costs — damage to human health, fish production, and the environment — we foot another $400 billion in expenses.

2. Impose a carbon fee on animal agriculture for the pollution their products cause. (See "Make Polluters Pay," page 236.) The rest of us pay to dump *our* garbage.

3. Make the USDA require

    *   technologies to capture methane from livestock burps and farts (cow backpacks are used by some ranchers in Argentina),
    *   biogas digesters to turn animal poop into usable energy,
    *   covered manure lagoons,
    *   composting, and
    *   feed additives to reduce farts and burps.

4. The USDA could also incentivize grass-feeding, which can sequester carbon (and corn makes for more gassy emissions). The USDA could also incentivize local farming and other forms of protein production. Iguana, for example, really *does* taste like chicken. And crickets have a much higher protein-yield per acre than beef, yet have wee little carbon footprints. Cricket flour is a booming industry (that's right: crickets ground into high-protein flour), but start-up costs are high for farmers.

Yes, plant-based diets are great for the planet and our health. But for a greater climate impact, get out the cattle prod for industry.

If you have...

Hungry kids: Crank up music by Formidable Vegetable Sound System (http://music.formidablevegetable.com.au), such as the album *Grow Do It*, while kids cook from cookbooks like *Pretend Soup* by Mollie Katzen.

A school open to change: Share Oakland's Unified School District's report on their successful low-carbon, healthy school lunches, "Shrinking the Carbon and Water Footprint of School Food: A Recipe for Combating Climate Change" (https://foe.org/projects/food-and-technology/good -food-healthy-planet/school-food-footprint). With other parents, push your district for a similar program.

Try "Meatless Mondays."

Time: Visit the National Sustainable Agriculture Coalition's "Take Action" page (http://sustainableagriculture.net/take-action). Take one action.

With your kids, watch "Happy Hangman," a fun two-minute video from the Reducetarian Foundation (https://reducetarian.org/resources).

Support pregnant or nursing moms who need or want animal products (or anyone who may need some animal products to address health needs).

With kids, read the young readers edition of Michael Pollan's *The Omnivore's Dilemma*.

10. Turn Fido into a Climate Hero

The average carbon paw print of our dog or cat is higher than an average human from countries like Haiti or Afghanistan.
— LARRY SCHWARTZ, SALON

When I was growing up, my family had a menagerie that ranged from gerbils to Welsh ponies, with the occasional piranha thrown in. Mostly, though, we were Dog People. Purebred German shepherds kept showing up because my sister befriended a dog rescuer who knew DeMockers couldn't refuse a lonely shepherd in need of kids and a yard. This is how, at fifteen, I came home from camp one summer to find Linda awaiting me.

Suddenly, I had a built-in best friend. Linda was a patient listener, uncomplaining jogging partner, and foot warmer who greeted me with joy the moment I opened my eyes each morning. She offered the unconditional love that drives countless North Americans every year to impulsively bring home big-eyed puppies. Dogs are good for the human spirit. Or, as my friend Tim says, "Dogs and people — we go way back."

The only problem with this interspecies lovefest, climate-wise, is dog food — or any pet food, really. Some use palm oil, which is a ruinous ingredient for our climate (see "Give Palm Oil the Back of Your Hand," page 19) or palm oil derivatives. And pet foods collectively use a lot of carbon-intensive beef.

If you have a pet, consider some of these solutions — though always pay attention to your pet, and consult your vet, to meet your lovable fur-ball's particular needs. Many dog-lovers buy palm oil–free food made from soy or a variety of other meats (which have less environmental impact than beef). Also, you might buy meat by-products from local butchers, which reduces pakaging and transport. For our dogs, we picked up a monthly supply of frozen and bloody cubes to toss into the basement freezer (which made my every trip into the dark, underground room for Linda's food feel a bit like a scene from a horror movie).

Cats are obligate carnivores who usually do well with poultry, lamb, pork, or fish; some cat-lovers request scraps at the fish market. While some owners advocate vegan cat food, the Feline Nutrition Foundation cautions that cats are very dependent on protein, and if they don't get enough "to supply their energy needs, they will break down their own body muscle and organs."

As with human food, it helps to buy organic food in recycled/ recyclable packaging, though that's pricier. Also make sure you aren't overfeeding your pet. As with its people, North America's pets now suffer increasing rates of obesity.

Then there's poop: Much of what is left on the ground in parks ends up in rivers and streams, wreaking havoc on water quality and wildlife. If you have dogs, pick up and, if possible, flush bacteria-laden feces in the toilet, so they're appropriately processed, not landfilled. To protect waterways, walk dogs elsewhere.

Check out Flush Puppies, which boasts a dog poop bag that is compostable (in industrial waste facilities that accept pet waste) or flushable in your home toilet (https://flushpuppies.com).

If you have cats, *don't* flush cat poop or kitty litter. Cats often carry a parasite that evades sewage treatment plants and harms sea creatures. Instead, try litter from sawmill scrap or newspaper clippings, which lets kitty share her opinion on articles that deny the reality of global warming.

If you have...

A cat: Keep kitty indoors to lower the number of birds killed annually by cats (some estimate one *billion* birds). If you let kitty roam, try CatBibs (exactly what they sound like — bibs for cats), which work better than bells to reduce bird killings; visit Cat Goods (https://catgoods.com).

A dog: To learn more about organic and palm oil–free dog foods, read the in-depth report by Ethical Consumer (www.ethicalconsumer.org /buyersguides/food/dogfood.aspx).

Make simple, nontoxic toys with rope, socks, balls, and toilet paper cardboard, or reuse your kids' outgrown stuffies. Find fun ideas online.

Try bamboo collars and leashes; natural, renewable bamboo doesn't contain toxic petrochemicals.

A hankering for a pet: Consider adopting, which helps with pet overpopulation. Then, spay or neuter your pet to prevent unwanted litters.

11. Cut Your Toilette in Half

Don't waste a single day of your life being at war with your body. Just embrace it.
— TARYN BRUMFITT, to her three-year-old daughter

W hen I was in middle school, the "hate the way you look" industry commandeered my brain. Within weeks of my first fashion magazine subscription, I dutifully traded my babysitting earnings for countless products, then rolled up my sleeves to correct my "beauty flaws." I shaved, plucked, permed, painted, deodorized, peeled, exfoliated, pierced, waxed, and body sculpted (big bosom! flat belly!) — and that was all before I applied the first coat of foundation.

I never achieved goddess-dom, but I got several eye infections from mascara. In college, I took my doctor's advice and, cold turkey, dumped my cosmetics.

Life became simpler, and I later married a man who prefers me au naturel. While parenting our daughter, Zannie, we emphasized wellness over beauty. We barred Barbie and her impossible body proportions from our home. When researchers publicized that all women feel badly about themselves within three minutes of flipping through fashion magazines, we steered Zannie toward healthy media. We also educated her about the ugly side of an unregulated multibillion-dollar industry, with its animal testing, toxins, and stealth attacks on self-esteem, which can contribute to depression, self-harm, eating disorders, and suicide.

Then, just as Art and I patted ourselves on the back for raising a comfy-in-her-own-skin girl to adulthood, I noticed a blotchy middle-ager in my mirror.

Once again, I feel the pull to paint and pluck. But I also know industry's beauty "remedies" don't just pollute through toxic ingredients, packaging, and transport. They also waste money and, more insidiously, life energy.

How do we break free and feel content with ourselves? How do we stop trying to be sex goddesses and become true goddesses — the kind that save planets and fight for justice?

My save-time-money-planet solution is to follow a three-part "feel-beautiful-enough" routine. Part 1: Get enough sleep. Part 2: Follow a simple routine that helps me feel confident. Often, in rainy Oregon, that includes blow-drying my hair. On date nights, I pull out black boots and red lipstick (and sometimes stray chin hairs). Part 3: Say "I love you" and "Thank you, body," to the mirror whenever I think of it. And usually have a good laugh with myself. This is the cheapest, simplest, and best way I've found to transform into a goddess.

Self-hatred is another form of dirty energy. The most powerful resistance to it is self-love. Which, by the way, is renewable.

If you want to . . .

Be informed: Research and share the benefits of body hair. Teens feel social pressure to remove or trim even pubic hair, despite higher incidences of sexually transmitted diseases in even moderate grooming (researchers surmise that hair removal can create micro-tears in skin, which are vulnerable to bacteria and viruses).

Watch *Story of Cosmetics* by Annie Leonard at The Story of Stuff Project (http://storyofstuff.org/movies/story-of-cosmetics) and Tayrn Brumfitt's documentary *Embrace* (ages twelve and up). Brumfitt also founded the Body Image Movement (https://bodyimagemovement.com), after a photo of her nude postpartum body rocked the internet.

Use healthier cosmetics: Buy fair-trade, organic, cruelty-free cosmetics and hair and skin products.

Or make your own. Find recipes online for homemade body-care products. Our kids loved making soap, bath salts, and lip balms. Use grocery items like coconut oil for face and body moisturizer, lip balm, and eye-makeup remover.

Cut your toilette in half: Challenge teens and yourself to shorten the time spent washing and grooming daily by getting rid of half of your "beauty" products. Whoever ditches the most gets a coupon for a treat, such as a massage, aromatherapy oils, or outdoor gear.

Raise body-positive girls: Buy no-name dolls at garage sales. Or make your own. A Child's Dream (https://achildsdream.com/doll-making) has instructions.

Discourage mall hangouts. Instead, steer girls toward the park, ice rink, or pool.

Dump fashion and gossip magazines, and stock *Kazoo* ("for girls who aren't afraid to make some noise"), *New Moon*, *Skipping Stones*, or *Ranger Rick*. For teens, try the online magazine Rookie (http://www.rookie mag.com).

Be a model — for loving your body the way it is and caring well for it.

12. Shower Bikers with Praise (Not Puddles)

Every time I see an adult on a bicycle, I no longer despair for the future of the human race.
— H. G. WELLS

It was a dark and stormy morning. Leaves clogged storm drains, transforming gutter rivulets into lakes. I drove slowly to the grocery store, windshield wipers energetically whomping back and forth. My preschoolers, inspired by the rhythm, kicked their feet frenetically on the back of the front seats, helpless with laughter.

Up ahead, a pedestrian walked the sidewalk, head down in the rain. An SUV in front of me sent up an arcing sheet of water right over the poor guy.

Right then, I made a silent vow: For the rest of my driving life, I would treat every human being using their own power for transportation — on foot, bike, wheelchair, skateboard, Rollerblade, or go-cart — as a demigod.

To the kicking kids, I said, "Let's be extra nice to anybody walking or biking."

"Okay!" they sang back.

Pedestrians and cyclists aren't polluting. They are out braving the elements, doing their part to *not* cook the planet. They are people like my husband, Art, who commutes by bike, and like my kids, who have hopped on bikes daily almost all their lives. Cyclists should be treated with great respect. They deserve dedicated bike paths, covered parking, workplace and school locker rooms, showers, and fresh towels.

It's not rocket science to make cities bike-walk-roll-centric, and many cities like Copenhagen promote the enjoyable, low-carbon movement of people. We just have to pay for it. And until the happy day when most people get around without fossil fuels, we can honor our walk-bike-roll climate heroes by how we drive (or, better yet, join them).

Supporting and honoring bikers and walkers teaches children to do the same. Researchers have found that kindness makes everyone feel good — including the giver, the recipient, and (I just love this) anyone *watching*. Yielding to a mom who is pushing a stroller — maybe even throwing in a friendly wave — makes our kids feel good just because they witnessed it. Taking those few seconds of extra time might even help us feel more calm and connected ourselves. And who knows? The pedestrian we protect today may become the driver respecting our child in the bike lane tomorrow.

Here are some easy-to-adopt habits to feel happier, model safety for our kids, and promote human-powered transportation.

If you are...

Driving on the road:
1. Pause for children crossing the street. Wave them across. Imagine they're yours and think protective thoughts.
2. When driving a hill next to a panting biker, stay in low gear to minimize exhaust. Flash a thumbs-ups of support. Pass cautiously; give a wide berth. Let kid passengers in your car know what you're doing and why.
3. When possible and it's safe, yield to bikers, especially if it's raining, windy, snowing, or hot. After all, you're in a comfy, temperature-controlled environment.
4. Be patient when waiting for pedestrians and bikers to cross in front of you. Instead of wishing they'd hurry, mentally thank them. And wait to turn or continue until they're safely out of the road.
5. Pay attention when making right turns and don't cut off bikers (they'll run into you)!

Parked or waiting off the road:
1. Turn the engine off. Don't idle. Politely request this of other drivers with a warm smile.
2. When opening your car door into a bike lane, use the "Dutch Reach": Dutch drivers open doors with their right arms, which forces the head to turn, making it easy to spot cyclists. This reduces "doorings," a common and sometimes fatal injury for cyclists everywhere.
3. Thank cyclists. Art says, "When strangers call out 'Thank you!' or 'Good for you,' it motivates me."

At city hall: Until we're fossil-free, push for protection of those who lead our way. Demand traffic calming — speed humps, midstreet crosswalks, protected crossings — and pedestrian malls, murals and gardens in intersections, and every form of flashing light or protection at crosswalks on busy streets. Demand bicycle infrastructure such as dedicated bike paths, bike lanes, bridges, and tunnels.

13. Swoon over Family Biking

The bike movement has grown up, and now it has kids!
— SHANE MACRHODES, Kidical Mass founder

The party wasn't quite over when my friend Mary grabbed her backpack and said to her three kids, "Sun's going down. Let's get home!" When the family rolled their various bikes from the garage, curious partygoers eating pie on the porch walked to the driveway and surrounded the family of cyclists.

Two moms inspected the tagalong that attached the six-year-old to Mary's bike while a couple of dads debated the merits of wider tires for in-town commuting. Everyone wanted to hear about gear, grocery-hauling, rain, hills, and what worked (or didn't). Mary's family commutes 80 percent of the time by bike, even through Oregon's soggy winters, and the kids proudly explained their gear like astronauts showing off spaceships.

The family biking movement has taken off in recent years and inspired countless innovations, such as the stylish kid and cargo carriers that make a biking life more comfortable and convenient for families. It's easy to fall in love with cycling for everything from summertime weekend recreation to year-round daily commuting.

Parents wanting to dip a toe into family cycling can join rides like the ones sponsored by Kidical Mass, whose motto is "Kids are traffic, too!" Families gather at a park for a short ride to another fun spot like an ice cream shop or local festival. The rides are a party on wheels and a rolling parade for the dizzying array of biking options — handlebar seats, trailers, Xtracycles, longtails, Long Johns, Bakfiets, tandems, folders, trikes, tagalongs, couplers, balance bikes, bikes with training wheels, recumbents, unicycles, cargo bikes, ellipticals, electric bikes, and bike tows with pedals for kids.

The climate-friendly benefit of cycling is secondary for many families, who love all the other things bikes offer: exercise, self-reliance, financial savings, easy parking, fun exploring, and a powerful mood boost.

One of the beautiful things about parenting is that young kids love what we love. If we swoon over monster trucks, so will they. Art and I swoon over waterproof panniers that keep books and laptops dry. Our son, raised to bike to school, doesn't yet dream of reflective raincoats and helmet covers, but he does, at seventeen, mountain bike for fun.

Here are some easy ways for parents to give biking a whirl.

If you want to . . .

Bike as a family: Visit Kidical Mass (www.kidicalmass.org) for gear guides, family cycling blogs, and links to chapters. Check out the annual Kidical Mass Ride, which happens in dozens of cities nationwide.

Check out the Family Adventure Project (www.familyadventure project.org), with in-depth blogs on family cycling and gear tips.

Get a city bike map (sometimes free through local government websites). Search YouTube tutorials for using city bus bike racks.

Take a class in "Confident Cycling" through your Parks and Rec (kids usually take safety tips from teachers better than from us parents!) No class? Request one!

Be inspired: Google "Copenhagen cycling" to see inspiring bike-only bridges, innovative bike racks, and businesspeople on multilane bike highways.

Watch stunt biker Danny MacAskill's joyous "Wee Day Out" on YouTube (www.youtube.com/watch?v=K_7k3fnxPqo).

Spruce up your bikes: Get decorating! Set kids free with the paint and glue gun. Add multicolored rim lights for nighttime safety and festivity.

Improve biking in your town: Launch a Walk-Bike-Roll initiative at your school to prioritize safe and fun school commutes. Demand no-idle zones, especially by schools, and covered bicycle parking.

14. Keep the Clunker — Until You Can Afford Cleaner Transportation

T he Frisbee team gathered early in the school parking lot. One of several parents driving the carpool, I'd prepared all week for the three-day Seattle tournament, even making a crosstown trip for the over-packaged wasabi seaweeds and trail mix the boys wolfed down last time. I love cars brimming with teens, especially when they laugh over games I can sometimes coax them to play.

"Boys, find a car," said the coach. Mine was a 1996 Camry wagon we'd bought used for moments like this. Though I lusted after a long-range electric vehicle (EV), my wagon was reliable and cheap to maintain and insure, had seven seats and a rooftop gear carrier, and most important to me, was paid for. All of this forgave its peeling paint and dented door.

The boys flew into SUVs and minivans. Only my son stood by my car. Due to last-minute changes, the team had exactly enough seats — if I didn't go. After I said, "Go ahead," Forrest beelined to a minivan, and I waved as his van drove away. Suddenly, I found myself crying.

"I guess I'm Loser Mom," I said to another mother. "And I even bought decadent snacks."

"Not at all," she assured me. "Those big cars are just so comfortable for long trips." I drove Loser Car home, resenting its dented door, hard seats, and lack of legroom.

"I need a grown-up car," I announced to friends that afternoon, "and Art needs a pickup truck." When they sang the praises of low-interest rates, however, a quiet panic overtook me. Any loan — even the 1 percent interest loan offered from a friend selling her used EV — would derail our careful budget.

I needed a hard-core financial planner. Luckily, I didn't have to look beyond my dinner date. My husband drives a faded-red two-door 1983 Toyota Starlet we bought before kids. It is tiny, scrappy, gets great mileage, rarely needs repair, and takes him on remote forest roads. No one tries to steal "Scarlet."

Art reminded me of her multidimensionality: She was his contractor's rig for fifteen years. She hauled four middle-school soccer players plus backpacks. She still navigates washed-out backcountry roads — even, once, under a huge fallen tree — then serves as Art's kitchen counter. She's the Giving Tree on wheels.

"An EV and a truck would look better," he admitted, "but our cars work. I say no loans." I felt a rush of love. He's not just unashamed to live within our means. He's proud of it, and of Scarlet. Together, we affirmed our mission: Support our contribution to our kids' college education, prioritize health and connection over things, and keep me free to rabble-rouse for climate recovery.

We kept our ugly cars. The kids supported our decision. I find ways to connect with the Ultimate Frisbee team — and Forrest — by cheering at games, helping with fund-raisers, and hosting backyard hangouts.

We're also working toward getting a plug-in EV. Cradle to grave, they're half of the environmental impact of gas cars — and even less in regions where they can be powered by clean electric energy. They're also coming down in price. A lot of used Nissan Leafs and other EVs are available for cheap, especially in bigger cities, when they come off lease. Many forward-thinking local utilities are giving big subsidies to get people into EVs, and California even has income-based subsidies. We are itching to stop buying gasoline from fossil-fuel bullies, but we have to wait because we're doing this the old-fashioned way: by saving. We refuse to join the buy-on-credit spree that by 2016 had pushed auto debt to an all-time high of $1.2 trillion. Instead, month after month, a sum of money works its way from our checking to our savings account. It's not much yet — the next five years of college tuition take priority — but it makes me happy. And as a lover of wordplay, I get a chuckle every time I note the "Auto Pay" in our account book.

One day, those little deposits will add themselves up to a sum I can convince someone to take in exchange for their used electric beauty. Until then, when it's time to drive the carpool to the annual Seattle tournament, maybe I'll rent myself a roomy SUV.

If you have...

An old, paid-off car: Keep it. Maintain it well. Save up for a cleaner car, watching for used EVs or hybrids or, soon, used e-trucks and self-driving cars.

With children, discuss your transportation options and their pros, cons, and affordability. Organize a family decorating party and turn your clunker into an art car.

Interest in buying an EV: To compare the average emissions of EVs versus gas and hybrid cars, visit (with older kids) the Union of Concerned Scientists interactive website (http://www.ucsusa.org/clean-vehicles/electric-vehicles/ev-emissions-tool#.Wd4epiMrInW).

Buy used, if you can, to minimize producing more cars.

No car: Stay that way, if you can. See "Swoon over Family Biking" (page 32), and rent cars for trips. Try car-sharing or transportation services such as Zipcar, Lyft, Turo, or Uber. To compare these services, see the Clean Fleet Report (www.cleanfleetreport.com/best-car-sharing).

For more advice, read the Carfree with Kids blog (http://carfree cambridge.com).

15. Don't Scurry, Be Happy

In the happiest of our childhood memories, our parents were happy, too.
— ROBERT BRAULT

I loved the pioneer definition of a "holiday" in *Little House on the Prairie*: Everyone ditched chores to walk into town for Fourth of July fireworks, gossip, and games of tag.

In my 1970s childhood, vacation meant nine kids piled into our six-door Checker cab, Jayco camper in tow, to visit our grandparents in Florida. On the way, we flew Kleenex "ghosts" out the windows; and once there, we played beach tag. I remember plentiful laughter.

With globalization, family vacations can become truly grand, involving Cancun water parks, Kenyan safaris, Disney cruises, or Egyptian pyramids. The irony, of course, is that our well-intentioned efforts to give our kids the world on a platter requires us to flambé it along the way. Air travel — with its massive carbon footprint — only adds more fuel to the fire.

What can travel-loving families do?

The *New York Times* advises, "The biggest single thing individuals can do on their own is to take fewer airplane trips." Author Mary C. Wood proposes we stay earthbound for a couple of years to help level off emissions at this critical time. Likening the climate crisis to a growing debt, she told me, "The idea is to make a down payment now before our debt spirals out of control. If we wait too long, we may be past the tipping point."

Keep this in mind when you plan your family vacation. North America is packed with natural beauty and vibrant cities, which is why international visitors flock here. What other continent can boast the Grand Canyon and coastal redwoods, New Orleans and Broadway? Consider destinations reachable by car, train, or bus.

Or ask your kids. My husband and I were once surprised when our children said they'd rather vacation in Oregon tree houses connected by ziplines (with relaxed parents unworried about costs and the stress of flying) than in a Jamaican resort (with parents harried by the same).

They're not alone. When researchers asked children what they most wish for, the majority responded, "For my parents to be less stressed and tired."

Art and I strive to be less stressed and tired. We try to unplug, ditch to-do lists, and really play. We fly less often to visit relatives now — too much money, stress, and carbon — and stay connected through calls and video chats. We vacation closer to home and spend the money saved by not flying on other things. If budgets allow, we go to restaurant dinners — which frees everyone from cooking to play more games of tag.

If you want to . . .

Assess the climate cost of travel options: Use the TerraPass Carbon Footprint Calculator (www.terrapass.com/carbon-footprint-calculator). Read about the complex drive-versus-fly equation. Ultimately, jet fuel, especially at high altitude, does more short-term harm than car fuel, so driving is usually better.

Have a short stay-cation: With your kids, plan a two-day stay-cation, and walk, bike, and bus to local off-the-beaten-track places. If your kids were foreign tourists, where would they go? Visit Atlas Obscura (www.atlas obscura.com), which lists strange and interesting sights most people have never heard of. As toddlers, our kids loved making gravestone rubbings at a pioneer cemetery; later, fun excursions included our local art museum, with a stop at the gift shop and a café treat. Camp for free or cheap. Skip stones on a lake. Splurge on something special. Explore where you live.

Enjoy a weeklong family trip: Plot how far you can get, and what you can see, in your region by car, bus, or train within your timeframe. Look for events, like festivals, that would make memorable experiences. Consider "slow travel": staying longer in one place to get to know it and the people more deeply. Or make the journey the destination and improvise as you go.

Consider renting — or buying used — an RV. These maligned gas guzzlers are usually cheaper and, used judiciously, less polluting than the typical flight/car/hotel combo. When our kids were toddlers, our $1,200 Toyota Chinook pop-up camper gave us many cheap, easy-on-Mom camping trips within an eighty-mile radius.

16. Bring Paris to You

Reading is a discount ticket to everywhere.
— MARY SCHMICH

One summer, when my daughter was a teen, two of her friends flew to Paris with their moms for their sweet-sixteen birthdays. We waved them off with a cheery "Bon voyage!" but felt green with envy and worried for the climate.

The carbon footprint of flying is a vexing issue for climate-conscious families who delight in traveling and engaging with other cultures. *J'aime Paris* as much as the next gal, but here on the West Coast, we're closer to Mongolia than Paris. According to the online TerraPass Carbon Footprint calculator, a round-trip for two from where we live to tag the Eiffel Tower spews 6.75 tons of carbon.

To visualize that, imagine standing on Manhattan's Sixth Avenue for the annual Macy's Thanksgiving Day Parade. After a dozen giant balloons of popular cartoon and movie characters float by overhead, along comes the grand finale: Santa Claus, rotund bellied and four stories tall. Children gasp in wonder at this huge floating apparition. Our round-trip Paris fling would fill eight of those giant Santas with carbon dioxide, which would join the billions of other Santa balloons–worth already in the atmosphere. About 55 percent of carbon dioxide pollution — from work trips, reunions, and vacays — eventually falls into the oceans, acidifying the water, which can dissolve shellfish and corals. The other 45 percent keeps orbiting and trapping heat for another hundred years, until our *great-great*-grandchildren turn sweet sixteen. Worse, scientists are finding that emissions from planes, since they fly at high altitudes, trigger a secondary set of chemical reactions that intensifies the warming. The Intergovernmental Panel on Climate Change estimates that the impact of airplanes is *two to four times* greater than the effect of their carbon-dioxide emissions alone. Air travel is particularly nasty on the atmosphere — and it's projected to double globally in the next twenty years.

What can climate-conscious Francophiles do? We could remain earthbound and jet-set less often. We could fly less far, maybe visiting

French-speaking Montreal (only four Santas for two round-trip passengers).

That week, Zannie's and my solution was to bring Paris to us. We splurged on clothes at downtown thrift stores, dined at a French restaurant — crepes and French onion soup — while reveling in French café music, and watched the movie *Amélie*. It wasn't exactly strolling through the Louvre (which you can do online), but it was fun.

This isn't just about alternative ways to celebrate our children's milestones. It's about avoiding or minimizing airplane flights whenever possible. It's about having exciting, even extravagant vacations that don't also trash the climate. See "Support Local Communities When Traveling Globally" (page 41) for more considerations, but one of the least-stressful, least-polluting, and least-expensive ways of exploring and understanding our world is to bring it to us. The *beaucoup* bucks of a Paris fling could fund a summer camp, a French class at the local college, or a visit to your nearest city's Little Italy or Chinatown. Don't care for Paris? Feel free to substitute Tokyo, Rio, or anyplace your young traveler is eager to go.

If you want...

To go to France: Try what Zannie and I did: Spend a weekend enjoying French cuisine and music (like *French Dreamland* by Putumayo), touring Notre Dame and the Louvre online, and watching French movies like *The Fairy*, *A Cat in Paris*, or *The Intouchables*. Read the *Adventures of Tintin* in the original French.

An international student to live with you: Host an exchange student. Exchange students are going to fly anyway; be their host family, and you'll learn about another culture. They might enjoy life with a climate-conscious family. Explore hosting an environmental studies exchange student.

To "offset" your impact when you fly: Instead of buying carbon offsets — criticized for commodifying pollution without actually driving emissions reductions — donate 10 percent of your flight's cost to grassroots groups working to keep dirty energy in the ground. Prefer funding clean energy or planting trees? Sure. Just do *something*, which is better than doing nothing, and models awareness of flying's impact for our children.

17. Support Local Communities When Traveling Globally

Leave only bubbles.
— CORAL REEF ALLIANCE

We've taken our children abroad twice in two decades. The first time, in 2002, was an all-inclusive vacation package generously gifted by relatives who didn't anticipate its colonial air. The walls enclosing the beach resort spanned the beach all the way to the water to keep locals out of the all-you-can-eat-and-drink, 24/7 beach party, and the dark-skinned Dominicans who babysat, taught aerobics, and served us poolside piña coladas were referred to by the resort as our "Chocolate Friends." (No, I'm not kidding.)

Years later, when we wanted our kids to hone their Spanish-speaking skills and begin cultivating a global perspective, we traveled down the Yucatán Peninsula through Belize and into Guatemala. That time, we avoided all-inclusive resorts in favor of smaller inns, and that helped us connect to people more easily, including to a nurse who invited us to visit the medical clinic she ran with volunteers in rural Guatemala. That day's visit ignited Zannie's desire to develop her Spanish skills and more deeply understand other cultures; she works now to support immigrant and refugee students in the US school system.

Families wanting something other than the standard two-week Hawaiian vacation have many options for helping kids become global citizens while engaging with local residents, particularly around climate impacts.

To start with, choose locally owned inns, restaurants, and transportation services instead of all-inclusive packages at big-box resorts, which mostly profit foreign investors. Or use a service like Airbnb, HomeAway, FlipKey, or VRBO to plan an immersive social experience, especially if your family shares space with a host.

For an active vacation that plunges kids into local nature and culture, try bicycling tours with kids — yes, even in other countries. Mother-of-three Kirstie Pelling writes on her Family Adventure Project website,

"Many people think you have to wait until the teenage years before trying something as ambitious as a multiday bike tour, but believe me, nothing is further from the truth. We've toured long distance with kids since they were babies."

To learn more about how global warming impacts residents in a particular region — especially the low-lying island nations that are popular vacation destinations — research before going, and then ask about local residents' experiences. Are they affected by drought, wildfires, floods, landslides, deforestation, seawater rise, habitat loss, or coral reef bleaching? Ask your waitress how her family weathers intensifying storms, or a doctor how the northward expansion of disease affects children she treats. Ask the grocery clerk, if you buy bottled water, what locals drink. Is it safe from contamination? Seawater encroachment? Talk to guides, bus drivers, museum directors. Befriend people. Really listen to responses. Ask how to help.

Dedicating even one afternoon to the real concerns of local residents — whether by learning about their environmental struggles or volunteering time to plant trees, restore coral reefs, or even just pick up trash — can forge authentic connections. It can also deepen children's understanding of their privilege and the effects of their actions and lifestyle choices — and of the impacts of their government's policies — on the families they meet.

If you want to...

Have an active family adventure: Peruse the Family Adventure Project's countless ideas for educational, active vacations with kids, including bike tours (www.familyadventureproject.org). (Caveat: They work with advertisers, so product endorsements are embedded in some posts.)

Experience coral reefs: Learn and practice coral reef etiquette at Coral Reef Alliance's website (https://coral.org/what-you-can-do/take-action/when-traveling).

Try service-minded travel: Find reputable local nonprofit organizations to support by visiting the Grassroots Volunteering database (http://grassrootsvolunteering.org) run by Shannon O'Donnell, author of *The Volunteer Traveler's Handbook*. Her Volunteer Opportunities database (http://grassrootsvolunteering.org/volunteer_opportunities) helps travelers find ethical and sustainable service-minded projects. Her Social Enterprises database (http://grassrootsvolunteering.org/businesses) helps travelers find, patronize, or donate to local businesses worldwide that have an underlying social mission, such as protecting the planet or empowering local communities (or both).

18. Get Electrified

*Gas extracted from shale deposits is not a "bridge" to a renewable energy future
— it's a gangplank to more warming.*
— ANTHONY R. INGRAFFEA, oil and gas engineer

When Art and I settled into our first house to start a family, we switched to "natural" gas for our stove, water heater, and dryer. "Natural" gas was cheaper than electric, easier to control when cooking, and the cleaner-burning fuel. Or so we were told.

True, burning gas produces less carbon dioxide than burning coal or oil, but there's more to the story: It's only cleaner if you burn it all. But released, unburned methane, in its first twenty years as a fugitive, traps *eighty-six times* more heat in the atmosphere than carbon. Yikes. Natural gas is like those spirits in the last scene of *Raiders of the Lost Ark* that seemed pretty benign until they started melting people's heads.

Mother Nature probably keeps "natural" gas deep underground because you just can't prevent climate-killing leaks when you drill for, pump, transport, liquefy, transport again by train or ship, re-gasify, and then distribute an invisible, odorless gas through countless pipes under countless streets into billions of appliances worldwide that may or may not work properly. This means that "natural" gas — particularly from fracking, which poisons water and then disposes of it deep underground — isn't the bridge fuel industry touts it as.

It's not only dirty energy, but according to a 2014 study, "worse than coal and worse than oil." For families, that means abandoning gas appliances and going full-on electric is definitely the cleanest choice. Over time, electricity will also get cleaner as coal plants are shuttered and utilities bring more wind and solar power online. Right now, switching from gas to efficient electric requires a pretty penny upfront, but for those who can afford it, there are options. Climate-conscious families can go electric with household appliances small and large, which, I'm happy to note, just keep getting sleeker and more efficient as demand grows.

Induction stoves, for example, are 73 percent more efficient than gas or conventional electric, and they're safer around children (they heat cookware, not surfaces, so they don't burn little fingers). It's also instant heat, like gas, but without the baked-on crud. Who doesn't love a glass top you can wipe? We are currently saving for one (as I remind myself, "It'll save money in the long run").

We're also eyeing heat pumps, which use half the electricity of standard furnace or baseboard heaters. And solar panels, which have an additional incentive: The number-one indicator that someone will buy solar panels is if their neighbor buys them. *You* can be the Joneses the neighbors are trying to keep up with!

Naturally, solutions must address infrastructure, not just individual homes. We need our cities to insist that utilities go green and move toward 100-percent clean energy. Cities must adopt green public transportation, with clean and beautiful electric trains, trollies, and buses. Electric school and city buses are ideal, since they don't go long distances and frequently must idle. (When diesel buses idle, they fill the lungs of passengers, especially little riders, with pollution.) With our kids, let's advocate for, and push representatives to adopt, low- or no-emission vehicle programs in every community.

Imagine the day when our kids get to ride those clean, diesel-fume-free electric buses they lobby for now.

If you have…

$0: Visit Pinterest for an eyeful of home solar and wind power, including an array of sculpture-like small wind turbines to inspire young engineers and designers.

Campaign against fracking and all new liquefied natural gas infrastructure. Visit FracTracker (www.fractracker.org) to find the gas pipeline nearest you. Download their app and join the campaign.

Some money: Get an electric or induction stove, on-demand water heater, solar water-heater panels, or passive solar for your house. Ask your utility about the process, cost, and tax incentives. Look into leasing solar panels. At EnergySage, learn about federal solar tax credits available through 2021 (http://news.energysage.com/congress-extends-the-solar-tax-credit).

19. Sleep Tight

Tell me, what is it you plan to do with your one wild and precious life?
— MARY OLIVER

My husband fondly remembers the screen-free bunkroom he shared with three brothers. The four boys were born within five years of one another, and they bunked together until, one by one, they left for college. For privacy, they each had one big drawer and an ironclad rule: Nobody touches a brother's secret drawer.

Maybe they get along so well today because they learned to share tight quarters and respect those drawers. Or maybe because this setup meant their parents were relaxed, paying a manageable mortgage on a two-bedroom home instead of a five-bedroom house with a man cave for each boy. Relaxed parents often raise relaxed kids.

Desiring to be at ease, at home with kids, and financially sound, I studied frugality. I read *Your Money or Your Life* by Vicki Robin and Joe Dominguez, who discuss the concept of "life energy." Life energy is the time, attention, or labor we give in exchange for services and goods that support our earthly lives.

Art and I agreed that in exchange for our life energy — Art's carpentry and teaching and my harp instruction — we wanted a home, health, happiness, and ample time to poke around in Oregon's natural beauty. The best opportunity to forge a healthy financial path, aside from avoiding divorce, disease, and arrest, arises during house hunting. Because wages kept sputtering along while home prices rocketed into the stratosphere, we decided to fall in love only with small (affordable) homes.

Despite having a boy and girl to house, we bought a two-bedroom, 968-square-footer. So, like their uncles, our kids shared a bunkroom until Zannie turned fourteen, when she designed, budgeted for, and happily built a tiny attic room with her carpenter dad. She can only stand up in the middle, but she loves the room she built.

My family's willingness to live simply in close quarters has saved us tens of thousands of dollars over what other families spend on the higher mortgage, insurance, utility, and repair costs associated with the average 2,598-square-foot US home. Now and then, do I covet walk-in closets and guest rooms? Yup. Do I fantasize about a master bathroom? I'd be thrilled with even a second bathroom. But not enough to surrender my freedom or, more importantly, my time-consuming fight for climate justice.

One thing that helps us stay the course is to put our lifestyle into global perspective; we try not to compare ourselves with neighbors on the next block, but with neighbors on the next continent. By global standards, we're incredibly privileged to own a small home, yard, and garage.

When I occasionally envy our friend Julie's gorgeous home, with its vaulted ceilings and tall windows onto stunning views, my kids respond, "Our house is the best!" Their cheerful attitude helps me recommit to my mission: Live simply, try to save the world, and love up my kids before they move on to homes of their own.

If you want...

Inspiration for living small: Check out two blogs: the Tiny House Family (www.tinyhousefamily.com), about living well with families in small spaces, and Becoming Minimalist (www.becomingminimalist.com).

To declutter a small house: Make a small house run smoothly with regular, ruthless purges. First, read "Find a Deeper Spark of Joy" (page 55). Then, read *Simplicity Parenting* by Kim John Payne and Lisa M. Ross and *Clutterfree with Kids* by Joshua S. Becker.

To downsize: Join the Small Is Beautiful movement launched by E. F. Schumacher's classic book (of the same name). Rent or buy a smaller home in which kids share rooms. Shared small spaces can be gorgeous and efficient and fun. For proof, take a peek at Lloyd Kahn's book *Small Homes: The Right Size*.

Revive the bunk bed! (They double as great play structures.)

20. Revamp Gift-Giving

Probably the reason we all go so haywire at Christmas time with the endless
unrestrained and often silly buying of gifts is that we don't quite know how
to put our love into words.
— HARLAN MILLER

Dear parents:
Usually, your children send me letters, begging me to fulfill their wish
lists. But this year, it's my turn to write. The elves and I can no longer deny
it: Our winter wonderland is melting. The North Pole is going *green*.

And your children's letters have changed. Many still want skates and
smartphones, but they're also frightened. Children in Asia ask me, "Make
the smog go away so we can play outside!" or "Please bring rain — we
need clean drinking water!" North American children ask, "Why aren't
adults fixing our climate?"

Worse, some write that I — through my annual stoking of everyone's desires for Things — also stoke the fires of the factories that produce them, which pollutes our atmosphere and unleashes terrible climate
impacts. Kids accuse, "You're acidifying the ocean my family fishes!"
"You're making Grandma ill during heat waves!" "You're spreading
dengue fever through my country!"

Parents, I'm melting my own North Pole. I'm complicit in destroying
— irrevocably — the homes, health, and happiness of the very children
whose dreams I'd hoped to fulfill with plastic play castles and flashing
Xboxes. But I hear your children saying, loud and clear, "Don't give us
Things that hurt the world and other children. Give us a future!"

So, starting today: I'm going "green" myself. To slow the melting,
we'll end our system of one-way accountability. I'll still keep tabs on the
Naughty and Nice, but everyone goes on the list now — me, you, and
Earth's businesses and governments. You've got real Grinches in power,
keeping kids in one part of our world madly consuming while, across the
ocean, others suffer, all to enrich investors.

From now on, the Nice will receive *one* gift from me — battery-free,
Earth-friendly, and fair trade. I'll bring recycled toys, science kits,
sidewalk chalk, and even "pre-owned" costumes, books, and musical

instruments, plus more off-the-couch-and-out-the-door gifts like fishing poles, kites, jump ropes, and skateboards wrapped in yesterday's comics or festive, reusable fabric.

For the Naughty: I'm divesting from fossil fuels, so they'll receive "Shrink Your Carbon Footprint Now!" booklets. If subtlety fails, it'll be Rudolph droppings next year — sustainably harvested, of course.

Love,

Santa Claus

P.S. I'm enclosing this gift guide so you can go green, too.

If you want a gift for...

Anyone: Visit New Dream's fantastic online resource center for gift-giving that deepens connections (https://www.newdream.org/programs/beyond-consumerism/simplify-the-holidays). It includes a pledge to simplify, a free "catalog" of alternative gifts, an alternative registry for newlyweds and expectant parents, and a guidebook for holidays "with more joy and less stuff."

Relatives: Show love with homemade cookies, a massage, a coupon for gardening help, or a donation in the recipient's name to a local nonprofit. Assign kids to create songs, poems, and artwork to melt hearts.

Young children: Simple stuffed animals and dolls, a magnifying glass, swords (wooden or handmade), cat's cradle string, wooden blocks, jacks, hand-me-down toys, tea sets, pads of recycled paper, and colored pencils. Try Waldorf school stores (www.waldorfshop.net) for handmade wooden toys, felt toys, and natural art supplies.

Older kids to teens: Try gyroscopes, pocketknives, books galore, graphic novels, comics, juggling sticks and balls, sports equipment, foosball tables, magic trick books and supplies, magnets, marble towers, art supplies, crafting kits, knitting supplies, compasses, slingshots, arrow sets (carve a bow!), dartboards, Frisbees, card games (like Trigger!), board games (Catan), flashlights with rechargeable batteries.

Other times of year: Establish limits on birthday spending. In grades four to eight especially, kids often fret over expectations for expensive gifts. Ask other parents to uphold a five- to eight-dollar limit. Throughout the year, stock a box with garage-sale or thrift-store bargains. "Regift" a treasure your family's done enjoying.

21. Raise a Financial Guru

It takes as much energy to wish as it does to plan.
— ELEANOR ROOSEVELT

"Uh-oh," I whispered to Art, "she found the camp brochure."
Thirteen-year-old Zannie flipped through the glossy catalog I thought I'd buried in the recycling bin, with its photos of girls laughing on mountaintops, girls performing skits, girls singing around glowing campfires. I'd never seen her so smitten.

It was my fault. I'd reminisced happily for years about my own summer camp. But tuition was out of reach for our family, and for two years I'd hidden those catalogs.

"Mom, can I just see the numbers?" Zannie examined the camp costs, zeroed in on the word *scholarship*, and asked, "If I get one, can you pay anything toward the rest?"

"Is it important enough that you'd give up violin lessons? If so, we'll let you choose."

"Deal!" She grinned happily, sent off an application, got accepted, and then got a job at a gymnastics studio. For the next three years, she worked every birthday party she could to pay her portion for camp every summer. Now, the camp pays her to be a counselor.

For his pricey year-round soccer club, Forrest cut a similar deal; he got a modest scholarship and paid 20 percent of the year's tuition himself. When he coveted brand-name cleats (see "Give the Youth Sports Teams a Time-out," page 57), he paid the difference between the used ones we'd buy and his dream shoes. He managed this by saving gift money, dog-sitting wages, and tooth-fairy dollars.

It helps that he and Zannie have watched me budget meticulously — one benefit of a small home. They know I love bill paying and see it as an elaborate game: Can our family eat well, own a home, travel, avoid debt, save for retirement, and have time together — all on teacher/artist incomes?

From observing me, they've learned a lot about everyday finances, like how one medical procedure generates several bills, hotels cost more on weekends, and you can shorten your mortgage through automatic payments. We model disinterest in the American Dream and instead model the pursuit of our own dreams through reflection and planning ahead.

We also made family money rules:

1. No. Loans. Ever. We front them money in a store *if* they have money at home, but they repay us immediately. We withheld even a taste of credit, that gateway drug, and this firm policy helped us avoid many a toy-aisle meltdown.

2. No payment for household chores. They're part of life. We do, however, crank music and enjoy sparkling juices on cleaning day.

3. We pay an allowance so they learn money management (their age in dollars once monthly, minus half, which is auto-saved in a college fund).

4. We offer work at an age-appropriate pay. Overpaying children is a disservice that ill-prepares them for a real job. (Other employers won't inflate my kids' pay just because they're adorable.)

5. We don't force tithing. They've sometimes donated some of their toys to families living in shelters. We encourage social change activism more than philanthropy — giving time, not just money.

6. Insist teens save for the future. By seventeen, Forrest was saving 75 percent of the wages from his job.

And I'm happy to report, Forrest reviewed this list and said, "I'll probably do those things with my kids, too."

If you have...

Young kids: Open a savings account for college and for shorter-term savings (like for camps). Use a nice box for savings (not clear jars, which can tempt siblings). Peruse Mint's recommended websites for teaching kids about finances (www.mint.com/ultimate-resources-for-teaching -kids-about-money).

Middle schoolers: Teach thriftiness — and how to fix bikes, cars, tools, equipment. Help them budget for things they want.

High schoolers: Help them open an account at a credit union with a debit (not credit!) card. Help them avoid annual fees and minimum-balance charges. Plot ways for kids to graduate from college without debt. Visit the hilarious blog *Mr. Money Mustache* ("Financial Freedom through Badassity," www.mrmoneymustache.com).

22. Find a Deeper Spark of Joy

We must rapidly begin the transition from a "thing-oriented" society
to a "person-oriented" society.
— MARTIN LUTHER KING JR.

Marie Kondo's popular *The Life-Changing Magic of Tidying Up* is a strange, wonderful book. I bought it mostly out of fascination: How can a book about tidying be number one on bestseller lists? But the next thing I knew, I was reorganizing everything I owned, hoping for that KonMari magic.

Her premise: To create a smooth-running home, you must purge excess, keep only what "sparks joy," and organize your remaining treasures according to her fail-proof method. So I dutifully pondered and purged. I held sweaters up and awaited the thrill. It rarely came. Out went tired clothes, unread books, clunky lamps. Faded towels? Gone. Good god, why hadn't I done this years ago? Under the KonMari spell, I replaced last century's raincoat with three new beauties — one for camping, one for around town, one for date night.

The problem started when I came up short at the month's end and had to return two raincoats in order to buy food. That's when I realized what my tidying had sparked: rampant dissatisfaction. I was mentally purging everything in my physical world. Did I adore my seven-dollar garage sale desk? No! Gimme a sleek ergonomic workstation! Our secondhand furniture? Hate it all! I deserved beautiful artwork, bright new towels, cute shoes! It was intoxicating. I couldn't stop wanting.

Luckily, I got to the part where she nearly wept over the slime on the bottom of her shampoo bottles after a few showers.

The spell broke.

It isn't just that Kondo is obsessive, which she cheerfully admits to being. It isn't even that she had no partner or child when she wrote her book, so she had time to dry bottles after showering.

It's that she never mentions the planet. Or anything, really, about nature or our relationship with it, or how our personal spaces and choices affect the climate, or that, just maybe, many of us live with objects that don't spark the "thrill of joy" for good reason: We can't afford those that do or we try not to consume more than we need. If everyone everywhere fills their lives with new, thrilling stuff — and sends piles of lackluster items to the landfill, as her clients often do — our world will be trashed.

Of course, I aspire to the organized life Kondo promises and appreciate how she greets her house when arriving home and bonds with her socks, which stand up straighter in her drawers. This makes me love Kondo. Like her, I want order, beauty, and responsive socks.

But I prefer to let the old couch work another year and forgive it for not sparking joy. I think, in fact, this can tap an even deeper joy: Old couches hold us through naps, family game nights, and emotional breakdowns, and being paid for, they free us for sacred work. Like parenting, and working for a livable planet — and the universal health care Kondo doesn't have to fight for in Japan. So thank your old couch, bike, bed, bathrobe, and all the small and miraculous things — like pop-up umbrellas — that keep you warm, safe, clothed, and fed. These things allow us to experience the spark of joy in our work and relationships, instead of trying to get it from our stuff.

With all the time and money we'll save, we can take on the life-changing magic of tidying up the planet.

If you have...

Five minutes: Pitch magazines and catalogs that inspire consumer lust. Unsubscribe from everything. Use the Federal Trade Commission's "Stopping Unsolicited Mail, Phone Calls, and Email" guide (www .consumer.ftc.gov/articles/0262-stopping-unsolicited-mail-phone-calls -and-email).

Twenty-five minutes: With kids, watch *The Story of Stuff* (http://storyof stuff.org/movies/story-of-stuff). Stock the bathroom, den, or bedside with puzzle books, cooking magazines, poetry, and life-affirming periodicals about wildlife, nature, or art (instead of magazines and catalogs).

The rest of your life: Don't subscribe to lifestyle magazines. In stores, withhold your address, phone number, or email, even in exchange for tempting discounts. Coupons, rewards cards, and frequent-buyer punch cards are designed to keep you buying.

23. Give Youth Sports Teams a Time-Out

We parents have elevated sports over all else, but there's no soccer in our kids' future if we don't control climate disruption.
— MARY C. WOOD

Last weekend, our high school Ultimate Frisbee team — twenty-six teens and supporting adults — traveled in multiple cars to a tournament hundreds of miles away to play against twenty-three other teams. In addition to spewing carbon dioxide, we left small mountains of trash in our wake. At the hotel breakfast alone, each team tossed dozens of foam cups, plastic utensils, and paper bowls on our way to warm-ups.

Youth sports culture has many benefits — kids are moving, outdoors, and learning teamwork and sportsmanship. Both of our kids have played youth sports, including team sports, and they're glad they have. Yet it's also true that, as parents, we often find ourselves caught between membership in an all-or-nothing sports scene and planetary citizenship. Carbon-spewing, trash-generating tournaments help fan the flames of global warming.

Youth sports have other negative impacts. They can dominate family life, which is increasingly sacrificed to the practice schedule. That means less time for other rich, meaningful activities (spiritual community, music, family time), as well as for helping kids join the climate revolution.

In addition, teams can stoke an intense brand consciousness. We were once informed, for example, that ours was an Adidas soccer team, and that while the club "won't *require* your child to wear only Adidas brand," well, it would be appreciated. This led my fifth-grade son to online shopping as a pastime, for the first time, and to develop an Adidas obsession.

Though some coaches actively avoid a winner-takes-all attitude, it bleeds over from the larger world of commercial sports young athletes are encouraged to follow. And that attitude spills, however subtly, to all aspects of our lives. This doesn't help kids' futures, since cooperation — not competition — is what we need right now to help our quest for a just transition to a clean-energy economy.

Is there a path forward for sport-loving families who are also committed to a livable future for our children?

With two kid athletes, we have struggled with this question and found success using, at various times, almost all of the approaches below.

If you want to join youth sports and also save the planet...

1. Forgo organized team sports for as long as possible. Instead, facilitate outdoor free play.
2. Create or join a community of families who raise "free-range kids" (see "Let Kids Play with Knives," page 140). One mom offers donuts to neighborhood kids who come around for street basketball or capture the flag.
3. Avoid year-round, one-sport travel teams in favor of doing several seasonal sports or martial arts. The added benefit: Children develop all muscle groups and avoid repetitive-strain injuries.
4. Create — and coach — a mellow league for local kids, as two of my dad friends did. Their "Safe at Home" baseball league stayed local and focused less on competition and more on fun and skill-building.
5. Encourage your child's exploration of a variety of sports. Especially for young children, seek easygoing coaches and leagues. YMCA teams are usually affordable, more relaxed, and lighthearted. Ultimate Frisbee prioritizes fun and community between opposing teams.
6. Join the growing number of parents challenging draconian "You're out!" policies if teens prioritize a holiday, religious ceremony, family reunion — or climate event — over sports.
7. Question far-flung tournaments. Art and I once successfully pushed against the climate footprint (and expense) of a tournament across the country. Parents opted for a regional one.
8. Organize carpools, and fill every car.
9. Use reusable water bottles. Avoid buying sports drinks in plastic bottles. You can also try making your own sports drink; search for recipes online, or just add salt and citrus to kids' water or juice.
10. Don't attend every game. Now and then, absolve yourself of cheering duty, and use the time for climate action. Parents can tag-team responsibilities — while one parent cheers on the sidelines, the other works on the frontlines for the planet's long-term health.

24. Be the 3.5 Percent

We need in every bay and community a group of angelic troublemakers.
— BAYARD RUSTIN

The twenty-two European students listened attentively to my lecture. Each came from an underprivileged background and had won a scholarship to study American environmentalism in Oregon. First, I described various ways I've tried to "creatively disrupt" politics as usual and sound the climate alarm, and I showed them ways others have done it worldwide.

Then I asked, "What's going on in your climate movements?"

A student from the Ukraine volunteered, "Here in the US you have big, daring protests. In my country, people don't talk about climate, much less protest the coal mine near us."

"Yes," a student from Siberian Russia chimed in, "there aren't even climate groups to join. Even though things have changed politically in Eastern Europe, our parents don't protest. They don't know how because for the last fifty years they've been told what to do by the government."

That gave me pause. I assumed that these students could rally their communities, especially for something as critical as not wrecking the planet. Their comments highlighted my North American privilege.

In many nations, environmentalism isn't just marginalized but dangerous. In 2016 alone, two hundred environmentalists, mostly indigenous, were murdered worldwide, and those attacks are on the rise. Even though US politicians and leaders have at times called climate activists "flat-Earthers" or even "pure scum," we're not slaughtered in our beds, as Honduran environmental activist Berta Cáceres was.

That's not to say there's no violence in the United States. In 2016, hundreds of indigenous water protectors and their allies protesting the Dakota Access Pipeline at Standing Rock experienced violence and abuse from police and private security firms using concussion grenades, rubber bullets, water cannons, and attack dogs. But no one was murdered.

Talking to the European students, I realized we have an advantage in the United States: Our culture loves nonconformists, daredevils, and rebels. We reward upstarts and bet money on start-ups. We were founded, in part, on rebellion, and we pride ourselves on a fighting spirit that has spawned movements for civil, women's, and gay rights. American protests benefit from 240-plus years of positive regard for self-determination.

And here's something heartening: University of Denver political science professor Erica Chenoweth, coauthor of *Why Civil Resistance Works*, discovered that nonviolent movements are usually successful *if* they involve 3.5 percent of the population.

Let that sink in. Only 3.5 percent! That's far fewer than the 18 percent of people who believe they've seen ghosts.

Ready to join us?

Let's be clear: This movement isn't hard to join. We're not trying to violently overthrow an illegitimate government or give gerbils the right to vote. We're not even trying to get the sixteen weeks of paid maternity leave my Dutch friend Klaarje gets.

We're just trying to avert collective suicide. And in our current period of US resistance politics, protests are increasingly viewed by many people as not only important and admirable but patriotic.

If people — or even you — are reluctant to join the 3.5 percent, remember this: The face of activism has changed since the tumultuous 1960s protests. The climate justice movement in North America is more intergenerational, intersectional, high-tech, savvy, and family-friendly than ever. People understand what's at stake (everything) and that we must use our incredible power wisely now — or lose it.

If you have...

Five minutes: Join a grassroots climate justice group near you. The largest is 350.org (https://350.org). Or join the online campaign or local chapter of the League of Conservation Voters (www.lcv.org), US Climate Action Network (www.usclimatenetwork.org), Sierra Club (www.sierraclub.org), Climate Parents (www.climateparents.org), Stand.earth (https://www.stand.earth), Rainforest Action Network (www.ran.org), CREDO Action (https://credoaction.com), Moms Clean Air Force (http://www.momscleanairforce.org), or Rising Tide North America (http://risingtidenorthamerica.org).

Time to read: Naomi Klein's *This Changes Everything: Capitalism vs. The Climate* or Bernie Sanders's *Guide to Political Revolution*.

Two hours: Watch *This Changes Everything* by Avi Lewis or *How to Let Go of the World and Love All the Things Climate Can't Change* by Josh Fox.

25. Pay Attention

Stop! Look! Listen!
— What adults tell kids to do before crossing the street

We should have known better. Earlier that summer, we'd shuddered over the story of eleven Europeans who, while exploring Arizona's dry-as-a-bone slot canyons, had suddenly drowned. A thunderstorm unleashed rain over dry, hard desert ground far upstream, and the torrent rushed downhill, trapping the tourists in the suddenly flooded narrow caverns.

Still, when we saw thunderstorms that morning from atop a mountain pass on our way to Arizona's Sycamore Canyon, we didn't fret. They were *waaaay* over on the left side of the horizon. Art, eighteen-month-old Zannie, and I were headed miles away, waaaay over to the safe, sunshiny right.

After a hot hour-long hike to the canyon floor, we eagerly suited up for a dip in a placid, toddler-friendly creek. As I changed Zannie's diaper, Art, hooting happily, waded toward the middle to dive. Then he froze.

"What?" I asked.

Beelining to shore, my impossible-to-alarm husband said, "I think a flash flood's coming."

I scooped up Zannie. "Should we run?"

"Yeah," he said, snatching the diaper bag.

We ran, barefoot, as the roar grew behind us. We bolted for a path uphill, and seconds later, a brown wall of mud roiling with plants, rocks, and tree trunks swept past.

"How did you know?" I demanded.

"I heard wind from upstream, but then didn't *feel* it," he replied, shrugging, as if it were obvious.

It's a good thing my husband pays attention. When he couldn't feel the wind, he listened more intently. *If it isn't wind, could it be water?* Then he remembered the thunderstorm and realized we must be directly downstream.

Almost twenty years later, that event still unsettles me. If Art hadn't sensed the flood coming, or if he'd hesitated, we'd be dead and you wouldn't be reading this book.

In his book *Deep Survival: Who Lives, Who Dies, and Why*, Laurence Gonzales writes that paying attention is the number-one thing survivors do. They notice changes around them — a strange noise or whiff of smoke, the baby's sudden silence, snowfalls becoming blizzards, levees breaking — and then, trusting their perceptions, they swiftly alter their behavior to save everyone or improve their chances in the new reality.

As the wind of climate chaos blows ever louder — decimating crops, causing unprecedented heat waves, hammering island nations with super-typhoons, and making flash floods more common — we'd best heed the warnings, grab our diaper bags, and run.

The problem we face now, of course, is that the fossil-fuel industry tells us not to trust our eyes, ears, or the peer-reviewed science that for decades has signaled danger. When we insist on rushing our children to safety, its executives, lawyers, and PR spin doctors — along with our own government — block the exits, fighting every attempt to regulate greenhouse-gas emissions or redirect oil subsidies toward clean and renewable energy technologies.

Families attentive to what's really going on can trust our senses, acknowledge the crisis, and then do what we must to improve everyone's survival odds. We must sharpen our observational skills and stay alert to true solutions.

This can start with wilderness survival itself, which is all about reading nature's warning signs.

If you want to...

Be spellbound: Read and watch "Snow Fall: The Avalanche at Tunnel Creek" by John Branch, a 2012 *New York Times*'s multimedia account of a preventable avalanche disaster (http://www.nytimes.com/projects /2012/snow-fall/#/?part=tunnel-creek). This captivated my family with riveting lessons in trusting your instincts.

Other movies (and books) that make for dramatic viewing and great discussions include *Into Thin Air*, *Touching the Void*, *And I Alone Survived*, and *Apollo 13*. What are the perils, errors, and decisive moments in each?

Teach yourself: Read, with kids, *Ultimate Survival Guide for Kids* by Rob Colson. As more Americans have discovered recently, basic survival skills are important to hone as we face more hurricanes, floods, mudslides, power outages, and fires.

Learn from others: Volunteer for search-and-rescue programs with your teens.

Or enroll older children in wilderness survival classes, camps, or extended programs like NOLS (National Outdoor Leadership School, https://www.nols.edu/en) or Northwest Youth Corps (http://www.nw youthcorps.org/m).

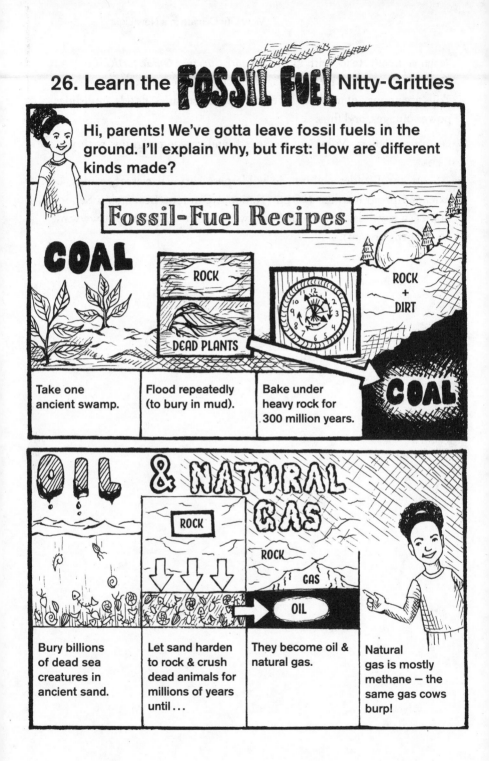

Hi, parents! We've gotta leave fossil fuels in the ground. I'll explain why, but first: How are different kinds made?

Fossil-Fuel Recipes

COAL

ROCK

DEAD PLANTS

ROCK + DIRT

COAL

| Take one ancient swamp. | Flood repeatedly (to bury in mud). | Bake under heavy rock for 300 million years. |

OIL & NATURAL GAS

ROCK

ROCK

GAS

OIL

| Bury billions of dead sea creatures in ancient sand. | Let sand harden to rock & crush dead animals for millions of years until... | They become oil & natural gas. | Natural gas is mostly methane — the same gas cows burp! |

27. Look to the Sky

It's time to look to the skies for the solutions that we need because the future of energy is no longer down a hole.
— Xiuhtezcatl Martinez

Ah, the sun. Ruler of the skies, giver of life — and provider of free, abundant, and clean energy. During our climate crisis, solar power is one of those proverbial rays of hope. This is partly because the solar-energy industry is exploding beyond every prediction — and how often do big transitions happen *faster and more cheaply* than anyone dreamed?

And yet, parents who are busy trying to keep kids off screens and on homework may not know that even cloudy Germany makes more than twice as much solar power as the United States does, or that 3.6 million homes in Bangladesh now sport rooftop panels. We might have missed that Morocco built the world's largest concentrated solar installation, or that China installs a soccer field–size set of panels every hour — and even built one solar farm in the shape of a panda.

This means that one way climate-conscious parents can help the solar revolution along is by touting its awesomeness. Solar energy costs are now as cheap as — and increasingly cheaper than — coal, oil, and even gas. And that's without subsidies and before tossing in the nasty health and environmental costs of fossil fuels. People *want* clean, affordable energy, and they're driving demand up and prices down.

But can fickle sunlight really replace fossil fuels? Not on its own, but it can be paired with renewable energies that are constant (in some places), such as wind, geothermal, wave, tidal, biogas from methane, and in-stream hydro. Forward-thinking utilities are implementing "grid flexibility," which means they can bring wind power on when the sun goes down, for example. It's kind of like symphony conductors cuing different musicians to play their individual parts. Other innovations include energy-use forecasting, as well as pricing that reduces demand at peak times, and ingenious new storage methods. Molten salt tanks, for example, are game-changers for solar-energy storage. Battery storage has also improved so much that in 2021 Los Angeles plans to unveil an eighteen-thousand-battery plant powered by wind at night and sun by day.

How can families and small businesses bask in the solar glory, too? Buying rooftop solar is a start, but many families face barriers because they rent, have shaded roofs, or can't afford expensive start-up costs. Third-party leasing, community solar, and loans offer some solutions. I'm mostly antiloan, but for solar panels, loans may be worthwhile because they can often be repaid quickly by directing the savings from lower utility bills to the loan itself. After the loan is paid, electricity is essentially free. Or try "net metering" — selling power back to your utility — all while cutting emissions. Making your own solar power is cheaper, cleaner, and quite literally empowering. It also makes our energy systems more diversified and resilient during storms, since whole neighborhoods aren't dependent on one centralized power source. Finally, it's an elegant form of resistance to big energy's stranglehold on our economy and democracy. Solar panels do produce greenhouse gases during production, as electric vehicles do, but both are emission-free once they're up and running. Cradle to grave, they're still far cleaner than any fossil fuel.

However, Big Oil isn't going down without a fight, which is why solar power — and all renewable energy — is in the Grand Oil Party's crosshairs. Some utilities are embracing renewables; some are doubling down on dirty energy. That's where advocacy is vital. Public utility commissions sound boring, but they're shaping our energy future. Let's make sure they keep seeing the (sun)light.

If you want to...

Join the solar revolution: Visit Solar Power Rocks (https://solarpower rocks.com/the-solar-blog) to explore all things solar and read reviews of everything from third-party leases to Tesla's solar roof tiles.

Explore alternative financing: Learn about community solar options at Shared Renewables (www.sharedrenewables.org).

Advocate locally: Push local utilities, commissions, and governments to subsidize solar installations for homeowners and to require or incentivize solar-power installations in new construction.

28. Be a WIMBY (Wind in My Backyard!)

Quite possibly one of the greatest discoveries hereafter to be made, will be the taming, and harnessing of [wind].
— ABRAHAM LINCOLN

In the early 2000s, a boy named William Kamkwamba was starving in Malawi. His family of subsistence farmers had watched in horror as drought withered their corn crop. Unable to attend school — no harvest meant neither food nor money for school fees — William became obsessed with a single idea: harnessing wind power to bring light to his house.

To learn how, he struggled through old science textbooks, scoured the local dump for parts, and experimented until, one day, he created a crude windmill that illuminated a light bulb.

Villagers who'd called him *misala* ("crazy") stared in wonder at his marvelous invention. William threw himself into building a windmill big enough to pump water for irrigation. Eventually, a large windmill turned above his home, bringing running water to his family farm and hope to his village.

Wind power. It's such an elegant — and egalitarian — form of energy. It's superior to coal, oil, and gas in almost every way: It requires almost no water and very little land; it can be up and running quickly; and once up, it peacefully coexists with grazing, farming, and recreation.

Wind has none of the hazards of burning fossil fuels: It doesn't spew greenhouse gases or poison aquifers and cause earthquakes, as fracking does. It doesn't spill oil into waterways and destroy fishing communities. It doesn't violate private-property rights and indigenous sovereignty, as tar sands operations and pipelines do. It doesn't explode. It doesn't require strip-mining mountaintops. It doesn't cause deadly respiratory diseases.

And wind, unlike fossil fuels, is unlimited. No one fights wars over wind.

Plus, wind power just keeps getting cheaper — in 2017, it was 2.9 cents per kilowatt hour, compared to 3.8 cents for natural gas. According to *Drawdown*, "The wind energy potential of just three states — Kansas, North Dakota, and Texas — would be sufficient to meet electricity demand from coast to coast."

That's the kind of US energy independence we enjoyed a century ago when windmills dotted our countryside — before fossil fuels took over. And if you're looking for job security for yourself — or your kids — the second fastest growing occupation in the United States is wind turbine service technician. (The first? Solar photovoltaic installer!)

Wind power does face challenges. It relies on mining rare Earth oxides for magnets and batteries; it doesn't enjoy the massive subsidies given to dirty energy; and it sometimes faces neighborhood resistance to new farms. Yet improvements in design, siting, and blade speed have reduced noise and bird deaths, a common concern. The greatest support for new wind farms occurs when: local communities are extensively involved in their development; farms are kept out of recreational areas; and ownership remains partly or fully within the community.

That's where families can help. We can encourage leaders to involve and empower local communities in new wind farm development. We can help our communities embrace wind turbines — instead of oil-drilling rigs — on our horizons, especially onshore where it's cheaper to build. We can tell the world that onshore wind production is one of the easiest ways to cut emissions quickly. Maybe we can even find ways, as cell-tower installers have done, to make them easier on the eye.

I love seeing wind farms because they're hopeful beacons of climate recovery. And for families in villages like Malawi, a single windmill can become a lifeline.

If you have...

$0: Borrow or check out from the library *The Boy Who Harnessed the Wind* by William Kamkwamba. Picture book, young reader, and adult versions are available.

Encourage WIMBY-ism in your community: Write a letter or commentary touting wind as an important path to our fossil-free future.

$14.95: Subscribe to the magazine *Home Power* (www.homepower.com) and learn how homeowners are making their own solar, wind, and hydro power.

$30 to $60 annually: Buy local wind from your utility. Many utilities have programs for supplementing home energy use with wind power.

An acre or more of land: Invite a full-size wind turbine to live with your family.

29. Be Here Now

Action is more important than prayer.
— THE DALAI LAMA

One day, while writing at a retreat center, I met a young Buddhist couple at breakfast. We chatted as we ate, and when we got around to my climate-action book and save-the-planet mission, the man began to politely spar with me. I was, he gently explained, causing myself unnecessary suffering and angst through my attachment to the world and catastrophizing over what might happen to the planet.

"I'm not attached to humanity's fate," he said. "And if whales go extinct, I'm okay with that."

"So, you have no children?" I asked.

"Right," he said.

I confess, I *am* attached. With every fiber of my being. I've even practiced "attachment parenting" because I think it helps children thrive. Does extending my love and compassion to all children, all future generations — through attachment *planeting* — make me some kind of drama queen?

Or is it, perhaps, creation's brilliant hardwiring to ensure that we're not all sitting in lotus position while the Earth burns? Every generation survives to adulthood precisely because parents' brains naturally imagine every possible danger lurking everywhere. It's why we don't give babies carrot rounds or let kids feed handfuls of blueberries to bear cubs.

My breakfast companion's gentle advice mirrors what I've received from many spiritual seekers over the years. They exhort me not to worry or "call in" negative energy. Instead, I should raise my vibration, relinquish illusions of control, change my thoughts, send love to bully presidents, and create my own reality.

"This moment," I'm told, "is all we have. Be here now."

Applied thoughtfully, it's sound advice. Applied too literally, it can be an elaborate shrugging off of both instinct and compassion. Because "be here now" means not only bringing yoga-mat awareness to our own breath and experience, but also to the suffering we may be causing others across the room, city, and world, not just now, but in the future.

Of course, let's focus on our breathing. It's good for our bodies and spirits and helps fortify us for the coming tasks. But let's also focus on the world's breathing. Because, yes, we're definitely creating, through our thoughts and actions — or inaction — our reality.

We're also creating realities for all present and future generations, for all time. With that kind of unprecedented power comes profound responsibility, and more and more spiritual seekers are gathering to broadcast just that message.

Perhaps the Buddhist Declaration on Climate Change states it most succinctly: "Future generations...have no voice to ask for our compassion, wisdom, and leadership. We must listen to their silence. We must be their voice, too, and act on their behalf."

If you have...

One minute: Breathe deeply and repeat one of these affirmations:

The Earth generously provides for my every need.
I have a sacred role in the world's unfolding.
I feel love and compassion for future generations and those with no voice in the climate crisis. I joyfully offer mine in service.
Today I will do one small thing to heal environmental injustice.

Eight minutes: Watch the Dalai Lama's appeal to the world to take climate action and care for our "only home," including for the Himalayas, the world's "third pole" that is key to global climate stability (www.youtube.com/watch?v=iYBMLsc64HM).

Thirty minutes: Take the last affirmation one step further to manifestation. Visualize yourself doing one small action, perhaps from this book, confident that mentally seeing yourself doing it increases the likelihood of actually achieving it. Then do it.

Time to read or listen: Read *Active Hope: How to Face the Mess We're in without Going Crazy* by Chris Johnstone and Joanna Macy, a Buddhist eco-philosopher and mother.

Read "The Fourteen Precepts of Engaged Buddhism" by Thich Nhat Hanh (www.lionsroar.com/the-fourteen-precepts-of-engaged-buddhism).

Visit Ecological Buddhism and read its Declaration on Climate Change (www.ecobuddhism.org/bcp/all_content/buddhist_declaration).

30. See the Human Face of Climate Impacts

Tell them we don't want to leave.
We've never wanted to leave.
And that we are nothing without our islands.
— KATHY JETÑIL-KIJINER, "Tell Them"

In 2015, countless residents of small island nations grew alarmed. World leaders were poised to gather for the December 2015 UN climate talks in Paris. At the time, this summit was widely considered humanity's last chance to hammer out a planet-saving climate-change pact. That was cause for celebration except for one thing: World leaders were set to propose allowing a global temperature increase of 2 degrees Celsius, even though scientists insisted that if temperatures rose more than 1.5 degrees, small island nations would drown. That extra .5 degree signified a death sentence for them — the people least responsible for melting polar ice caps.

As we are reminded every hurricane season — especially in 2017, when hurricanes Harvey, Irma, and Maria flooded Houston, obliterated Barbuda, and devastated Puerto Rico, respectively — coastal cities and, in particular, island nations are disproportionately punished by natural disasters, which will only increase as seas and temperatures rise. Small island nations are also often disproportionately poor, and according to the World Bank, "The poor — both those living in poverty and those just barely above the poverty line — are already the most at risk from climate change."

Just before the Paris summit, a coalition of nations launched the "1.5 to Stay Alive" campaign, to put a human face on the climate crisis and rally popular support for setting the temperature-increase limit at 1.5 degrees. Rather than keep waving the tired image of an iceberg-stranded polar bear as a symbol of global warming, islanders shared stories of real people — especially families — heavily impacted by rising temperatures. They hoped that people would respond generously, even heroically, if campaigners could inspire compassion.

To help bring more marginalized voices to the world summit, the web-based grassroots group Avaaz — with 43 million members world-wide — raised money to send delegates from the Marshall Islands, which is already being swallowed by rising seas.

The new approach electrified the summit. Marshallese foreign minister Tony de Brum told the world that small island nations lack the infrastructure and money to withstand intensifying storms. He announced a High Ambition Coalition of one hundred countries, whose goal was an ambitious, science-based, and legally binding agreement. Meanwhile, the "1.5 to stay alive" slogan was echoed by musicians, negotiators, Avaaz, and other global activists. Hundreds of youth chanted it in the streets, including those from Earth Guardians (see "Meet Some Badass Kids," page 115).

Kathy Jetñil-Kijiner, a young mother from the Marshall Islands, read her poem "Tell Them" to UN negotiators. Other Marshallese described seeing the remains of their loved ones washed away from flooded graveyards. To a listening world, the media spread stories of people whose homes and culture were vanishing.

And it worked. World leaders did something no one had considered a political possibility a month earlier: They formally agreed that small island nations have the right to remain livable, and they pledged to keep global temperature increases "well below" 2 degrees, with a target of 1.5 degrees.

People power works. And messages about how climate change affects people work, too, as was shown so effectively in Paris. Yes, we care about all creatures, including polar bears, but climate destruction is violence that hurts everyone, especially the world's most vulnerable peoples.

If you have...

$0: Visit the website of 1.5 to Stay Alive (www.1point5.info), which has news, information, and a "1.5 to Stay Alive" song.

Read more about Marshallese mother and climate activist Kathy Jetñil-Kijiner (www.kathyjetnilkijiner.com). Her website includes poems, videos, and messages about parenting, indigenous culture, and the effects of rising seas on her home.

$1 or more: Donate to Avaaz (www.avaaz.org), whose efforts put a human face on the growing climate crisis.

$500 or more: Like Avaaz did for the Paris summit, look into — and fund — bringing people from climate-ravaged areas so they can share their stories about the effects on their families. Connect with local or national chapters of worldwide environmental organizations, like Avaaz, Earth Guardians (www.earthguardians.org), Our Children's Trust (www.ourchildrens trust.org), 350.org, and Friends of the Earth International (http://www .foei.org).

31. Get the Facts

We are under siege by fake information that's being put forward by people who have a profit motive.
— ELLEN STOFAN, NASA's former chief scientist

When I moved to New York City in the 1980s, I loved the culture on commuter trains. I appreciated the quiet, library-like mood as we all swayed slightly back and forth. The majority of us read (this was before ear buds and smartphones), and I enjoyed trying to figure people out a bit based on their reading choice — the *Wall Street Journal* for the business-minded, *Vogue* for the fashion-conscious, and the *National Enquirer* for lovers of gossip disguised as news.

Fast-forward three decades, and newspapers and magazines have been replaced by phones, laptops, tablets, and earbuds. On the upside, the digital media explosion has allowed new perspectives, greater accessibility, and plentiful choices for engaging information. Long train rides go by in a flash. The downside is that it's harder than ever to discern credible information from opinion, propaganda, and outright lies disguised as news, partly because anybody can launch articles, films, websites, or even institutes. This collective confusion is poorly timed, though, given both our climate crisis and top political leaders who deny its existence. We need reliable information more than ever.

For decades, the Environmental Protection Agency provided stellar climate science on its website, but that science was scrubbed in 2017 under EPA administrator Scott Pruitt. Now, it's easier to stumble in our search for good information, and one place to be especially wary is on Wikipedia. The free, online encyclopedia "anyone can edit" is the seventh-most-visited website in the world, which means it enjoys a huge audience. Yet Wikipedia doesn't have enough editors to effectively vet sources for its millions of articles. This is problematic for many reasons, including Wikipedia's special vulnerability to "sock puppets" — people who use its huge platform to essentially advertise for paying clients by writing favorably on Wikipedia or by writing "articles" to be cited in an online entry.

Test this for yourself. For example, I recently checked every citation under Wikipedia's entry for "natural gas" — a subject that's increasingly controversial, given the public's growing awareness of its culpability in global warming — and I found half of the citations unreliable. They either led to error messages (10 percent), to citations not hyperlinked and therefore unverifiable for online research (20 percent), or to the ultimate pro-gas sock puppets — oil and gas industry websites, trade associations, and journals (20 percent).

Luckily, more reliable sources for climate information exist — just look to the people whose job is to get the facts straight: journalists. The Society of Environmental Journalists has compiled a great list of reliable environmental and science news sources.

In other words, if you or your young scholar need the facts on global warming, consider (and verify) your sources, and don't rely on public-participation sites like Wikipedia. And as your expertise grows, consider helping Wikipedia diversify its viewpoint — editors are 91 percent males under twenty-six. Editing's not for everyone, but if you have the facts, and Wikipedia doesn't, you can change that. Tech-savvy parents and teens can help Wikipedia present a more accurate story about, say, "natural" gas — by documenting California's twenty-one thousand known leaks, or by including the growing global resistance to fracking for gas. Somewhere, there's an entry awaiting your input!

If You Want...

Reliable climate news and information sites: Visit the Society of Environmental Journalists (www.sej.org/library/climate-change/staying-up-to-date-publications-follow), which lists trustworthy sources, like the *Daily Climate*, *ClimateWire*, *Reuters*, the *New York Times*, *RealClimate*, Yale Forum on Climate Change and the Media, and *InsideClimate News*, to name a few.

The climate science that was scrubbed from the EPA website: Visit the City of Chicago's website (http://climatechange.cityofchicago.org).

Fact-checks on general news: Try FactCheck.org (http://factcheck.org) and PolitiFact (www.politifact.com). OpenSecrets.org (www.opensecrets.org) reveals donations received by our representatives. Snopes (www.snopes.com) debunks urban legends (not for youngsters).

Tips for kid media literacy: Common Sense Media (www.commonsensemedia.org) and the Center for Media Literacy (www.medialit.org) both offer free media literacy advice.

CARE FOR
YOUR SOUL

32. Howl When Necessary

You care so much you feel as though you will bleed to death with the pain of it.
— J. K. ROWLING, *Harry Potter and the Order of the Phoenix*

I'm learning about grief from my friends Naomi and Jeff, who lost their twenty-three-year-old daughter to medical negligence. When Talia struggled to breathe after a neck fusion surgery, Jeff, who is a physician, feared that his daughter's airway could close off. He pleaded with staff to evaluate his daughter's airway and take steps to avoid an airway emergency, but his concerns fell on deaf ears. Then Talia's airway did close, and my friends watched as their terrified girl asphyxiated.

Devastated, my friends get out of bed every day to care for their other children. They teach me that parents don't get over that kind of trauma; they learn to survive on a minute-by-minute basis. For them, Talia's death was a careless killing, and nothing can change that.

My friends also teach me how celebration can coexist with body-slamming grief and up-all-night rage, how howls and hallelujahs intertwine as we mourn. Months after the burial, for example, Naomi discovered a sculpture of Talia's she'd never seen. Thrilled, she turned it over and over in her hands, touching where Talia had touched, delighting in her artistry, all while sobbing, wrecked anew by loss.

The global-warming crisis can also inspire grief over careless killing. It can inspire rage at negligent leaders who either willfully ignore reality or mouth blithe reassurances — even as the climate strains under its poison load and Earth's life-sustaining systems unhinge. As parents we can feel piercing, personal loss for the environmental tragedy our children must endure, yet we have no broken body to carry, keening, into the streets.

How do we mourn polar ice caps? How do we parent well and cheerfully when millions of children — as precious and beloved as our own — starve in lands baked hard by global warming? How do we go on when we know this careless killing is preventable?

In my family and community, when one of us reels from yet another oil spill or devastating wildfire, others listen, rub shoulders, pour tea. Sorrow flows faster when shared. Isn't this what people who struggle have always done? Share tears, stiff drinks, road trips, sacred ceremonies, or whatever transforms despair into power?

We also remind one another to cheer every triumph and bask in beauty. Everyone I know rejuvenates, paradoxically, in the places most threatened — in embattled old-growth forests, the acidifying Pacific Ocean, or on mountains once capped with snow. Like Naomi turning her daughter's sculpture over in her hands, we immerse ourselves in wilderness, cherishing and mourning it simultaneously, the howl and the hallelujah, that mash-up of heartbreak, solace, and resolve.

Afterward, our heartache becomes fuel for the fight. Naomi and Jeff now advocate for hospital reform so other parents won't have to bury their children as they buried Talia. We climate justice activists can also elevate our rage and sorrow to the public sphere — on the editorial page, the capitol steps, soccer sidelines, everywhere. We can join in the global chorus howling "No!," a cry that springs equally from our hallelujah, our beautiful "Yes!" shouted to a world making its decision.

If you are...

Alone: Heed your body and heart, and allow yourself any much-needed retreat, rest, or soul renewal. Nap. Stretch. Get off social media sites that exacerbate isolation and powerlessness. Listen to your favorite music.

And read. Here are books that can help:

Martha Whitmore Hickman, *Healing After Loss: Daily Meditations for Working Through Grief*

Kathleen Dean Moore, *Wild Comfort: The Solace of Nature*

M. Jackson, *While Glaciers Slept*

Francis Weller, *The Wild Edge of Sorrow: Rituals of Renewal and the Sacred Work of Grief*

Rebecca Solnit, *Hope in the Dark*

With others: Take turns listening. Hug, share soup. Ask your family for a "TLC Night," when a sad family member gets tender loving care — and a break from chores.

33. Change Your Media Diet

One reason that cats are happier than people is that they have no newspapers.
— GWENDOLYN BROOKS, *In the Mecca*

Raised with limited TV, I didn't own one as an adult. Art and I sometimes rented videos with our kids, but to watch our University of Oregon Ducks lose the Rose Bowl or an election get stolen, we had to invade the neighbor's den.

Then, Netflix and teens happened to our laptop. Our occasional videos yielded to the *Merlin* series, *Downton Abbey*, and finally *Friday Night Lights*. After season one, the characters invaded my brain. I'd wonder how Matt, the golden quarterback, would care alone for dementia-plagued Gramma. While I tried to write an inspiring invitation to a climate rally with real kids at an actual courthouse, I drifted into reveries about whether Lyla would get pregnant by Tim, the best friend of her now-paralyzed boyfriend, Jason.

One day, I blurted to my family, "I'm a soap-opera addict." We watched at night, but the plotlines mirrored the plots in daytime soaps. Unwatched episodes were as irresistible as warm cookies. I didn't visit friends as often or play music or attend pipeline hearings. Part of this was because it was a bonding activity with my kids, but mostly it was easier to just...watch.

I wasn't alone. My daughter accessed my account from college and admitted she binged because it was easy. When neighbor kids did the same, I realized six of us — five of them kids — were watching addictive shows thanks to my Netflix subscription.

So I canceled it. Even though I'd just discovered the series *Abstract: The Art of Design*.

Binge-watching is time-consuming, but also risky for climate justice because TV pacifies us. When we watch shows, we relax and our brain function slows. When we stop watching, the relaxation stops, but we continue to feel passive and dull. Perhaps that's why heavy watchers aren't as likely to partake in community activities. Apparently, the more we watch, the less we act.

Some shows can even imperil our health. TV news, with its breaking, graphic stories from anywhere, 24/7, changes how we see our world and our role in it. Researchers found that viewers of sensational news tend to exacerbate their own worries and suffer from more depression, anxiety, and symptoms of PTSD than people who skip the news or watch positive or neutral news.

Our kids can't afford depressed or pacified adults. They need us to wage climate revolution... My solution, now that Americans consume an average of ten-plus hours of media daily, is to *not* keep up. I stay decently informed — without TV. No CNN, Colbert, or a single series because, if I watched, I wouldn't be writing this book. When people insist, "You've *gotta* see this — I'll send the link!" I say, "Please don't."

Everybody's still speaking to me, so far, anyway. Even my kids, who, as they mature, must figure out what to watch on their own. That's okay. I've at least gained my own mental control back. And that's key — modeling empowered media usage.

I suggest we treat our news feed like food: too much and too low-quality renders us lethargic and disempowered. Try a media diet change, and see if you're not happier, as I am. If you're still not convinced, come over to our house and join our nightly Catan battles over land and resources — instead of over the remote.

If you have...

No self-control: Buy a router that limits internet usage. Mine cuts me off an hour before bed for better sleep. Delete news feeds from your phone and computers. Or try the app NPR One (www.npr.org/about/products /npr-one), which lets you personalize — and localize — your National Public Radio news stream.

A little self-control: Cancel cable TV, Netflix, Amazon, Hulu — all of it. Buy single movies or shows.

Try screen-free vacations. We negotiated this reunion policy when kids were young. All the relatives — teens, too — now cherish our unplugged week in the real world with swimming, hiking, and guitars around the campfire together.

A good amount of self-control: Try a family Sunday screen Sabbath at home, even for half a day.

34. Build a Wall — of Inspiration

A hero is no braver than an ordinary [hu]man, but [she or] he is braver five minutes longer.
— RALPH WALDO EMERSON

In her den, my friend Mimi has a ten-foot-long photo wall. You can find every member of her and her husband's families on it, solemn visages of long-dead relatives floating beside her children's baby pictures and snapshots of thirty laughing faces at a recent bar mitzvah. One glance as she breezes past with the laundry reaffirms for Mimi her belonging in a clan that adores her.

Art's and my far-flung families also smile from photo albums and fridge magnets, and our hearth displays baby announcements and artwork. This fuels my children's sense of belonging and makes me happy.

Some days, though, my save-the-planet work requires high-octane motivation just to swing my feet to the floor. One morning after President Trump's election, I felt drained and alone in my quest as I said yet again, "Hey, seriously, can we skip the mass suicide?" I realized I needed either amphetamines or a wall like Mimi's. Instead of relatives, though, I needed badasses. I needed a wall of freedom singers, truth tellers, and I-have-a-dreamers. I needed Gandhi politely defying the Brits. Harriet Tubman smuggling slaves. Susan B. Anthony voting before the boys said she could. Wangari Maathai illegally planting trees in Kenya until soldiers stopped arresting her and grabbed shovels.

I also craved connection with living heroes who fight fossil-fuel bullies, as I do, between packing lunches and helping with math — parents like Abby Brockway, who blocked explosive oil trains; my friend and kayaktivist Adriana Voss-Andreae, who helped block Shell Oil's icebreaker headed to the Arctic; or Sandra Steingraber, jailed for fifteen days because she stood between a reckless fossil-fuel project and her children's drinking water.

So, for Christmas 2016, I asked my family to create a wall of kindred spirits. My husband built a simple wooden altar, and as I write in my studio now, a host of parents cheers me on in photos. Every day, I keep company with those climate heroes as well as mother-of-seven suffragette

Elizabeth Cady Stanton, the Sweet Honey in the Rock singers, Bill McKibben, Honduran environmentalist Berta Cáceres, the Yes Men, Terry Tempest Williams, Winona LaDuke, Barbara Kingsolver, Alice Walker, Rebecca Solnit, and my friend Leonard Higgins, a grandfather who, with Gandhi-like peacefulness, keeps putting his body on the line to block dirty-energy projects.

When I'm ragged and without the strength to go on, my heroes silently say, *You've got this, dear. Keep on fighting.*

If you have . . .

Five minutes: Find a photo of one climate justice hero, or any hero. Print and tape it above your desk, or include it in your screen-saver gallery.

An hour and a tiny space: Create a mini-altar of loved ones and inspiring heroes. Get as crafty as you want. For pocket-size altars, Pinterest has thousands of mint-box altar ideas; glue an inspiring quote into the mint box alongside a photo, and place it, open, on a windowsill, shelf, or your desk. I travel with my mini-altar — a beeswax candle, matches, family photo, and inspiring quote or poem. For long writing retreats, I let the heroes from my altar stow away in my suitcase.

A free afternoon and more wall space: Make a wall of heroes, family, and friends. Use a bulletin or linen board, a corkboard attached to the wall, a bookshelf, or even plain old tape stuck to the wall. Add meaningful symbols — bumper stickers, notes from heroes, stones from special places, magazine articles about victories. You could also assign kids their own sections to fill — and change — as desired.

35. Get Your Soul Juice

A blessing is a circle of light drawn around a person to protect, heal, and strengthen.
— JOHN O'DONOHUE

We all need encouragement — sometimes even to rise from bed in the morning — and the right words can help banish despair, illuminate the miraculous, and reignite our ferocity for life. Every morning, before I let phones, email, or housework demand my attention, I say my own daily invocation (below) and read a poem or two to hearten me. Good poetry can feel like a homecoming, the way it hitches our attention to the sublime, and helps inspire us to act on behalf of all we love.

Morning Prayer

May I rise grateful for this day.
May I return the world's generosity.
May I love others and, above all, myself.
May I discern, but not judge.
May I remember that those who break others are themselves suffering in
   some way, but still do all I can to stop them.
May I channel my grief into effective actions.
May I seek spiritual refreshment regularly and help when I need it.
May I shelter my family and myself, always remembering that our journey
   together will be brief.

As we lie in bed at night, relaxing before we slip into dreams, many of us focus on our failures and process negative feelings. For climate-conscious parents, that can mean more than the usual parental tossings and turnings, since we also struggle with long-term anxiety over our children's safety. Lately, my husband and I have tried to wind down emotionally by moving what we fondly call our "10:30 tirades" — expressions of our shared outrage over world events — to earlier in the evening.

Then, before sleep, I often say a different invocation:

Nighttime Prayer

May I now rest, knowing I've loved well today.
May I reflect on this day, not in self-reproach but with curiosity
*   and kindness.*
May I thank my hardworking body with rest and replenishment.
May my family enjoy deep, peaceful sleep and good dreams.
May I be open to tomorrow's unimagined possibilities.
May I wake with restored hope, grateful for the new day.

If you have . . .

One minute: Put your favorite book of poems by your bed and read one daily upon opening your eyes. Notice how the habit affects your mood.

Fifteen minutes: Find a poetry collection to read from. Some poets to consider, if you're unsure where to start, are Mary Oliver, Anis Mojgani, Pablo Neruda, Maya Angelou, Wendell Berry, John O'Donohue, Hafiz, Jack Gilbert, and Naomi Shihab Nye.

Search online for dynamic young social change poets and performers, such as the duo known as Climbing PoeTree (http://www.climbing poetree.com).

Use any of the collections edited by Elizabeth Roberts and Elias Amidon, especially *Earth Prayers: 365 Prayers, Poems, and Invocations from Around the World*.

An hour: Write your own morning and evening poem/prayer/meditation. Poet John O'Donohue suggested that you write and memorize "a prayer that's worth the great destiny that's been given to you."

36. Try Protest Therapy

Usually when people are sad, they don't do anything. They just cry over their condition. But when they get angry, they bring about a change.
— MALCOLM X

"Breathe in! Let it out, fast," the workshop facilitator intoned. It was 1998, and ten of us lay on the floor, eyes closed for our weekly therapy session. We deep-breathed ourselves to a nearly altered state, then expressed out loud whatever the music "brought up" for us, sobbing our grief or screaming into pillows at bosses, partners, or even long-dead parents.

It was a noisy affair, but I usually felt grounded after all the hyperventilation, ready to work again toward peace within while spending my days with a toddler in the throes of her terrible twos.

But this time, the leader really cranked the music. All I could do, instead of relaxing, was yell, "It's too chaotic! Can't you turn it down?" The facilitator gently repeated into my ear, "The chaos is within you. It's not about the music. It's about your response to it."

I did my best, but the music *was* too loud. I should have said, "You're right, my response *is* in my control," and then stood up, walked across the room, and turned down the volume.

Famed psychologist James Hillman once commented that, when he was growing up, he remembered fathers going to political meetings to demand fair wages, workplace safety, and desegregation. Now, he pointed out, Americans are more likely to go to support groups, bonding over "pathology" — overeating, alcoholism, codependence.

"Suppose," he suggested, "we begin seeing ourselves not as patients but as citizens." He went on to question the way many of us use meditation as an escape, commenting that "the world is in a terrible, sad state, but all we're concerned with is trying to get ourselves in order."

I've used meditation and therapy and will again. In fact, a therapist who specializes in working with artists gave me critical support as I embarked on this book. Both resources serve important roles. But so does anger, especially now, when our government and financial institutions betray our families by investing in our collective destruction.

We don't have to choose between self-care and activism. Meditation and therapy can help us become more effective politically. But that's not what I see happening. I often want to beg friends and acquaintances to toss the self-help book and instead to help me topple tyranny, for heaven's sake. We aren't crazy to rage over the fact that global warming threatens everything. We're not mad to crave justice and peace. We *are*, perhaps, crazy for not jailing CEOs who knowingly destroy the Earth, or for not politically tar-and-feathering representatives who let CEOs write laws that wreck the planet (yes, the American Legislative Exchange Council literally writes up laws; see the endnotes on page 280).

In fact, these days, protesting seems like the only sane thing to do. And protests often work, especially as part of a major campaign. In 2017, protests against President Trump's proposed refugee ban targeting Muslims had a huge impact. People mobbed airports in an unprecedented uprising, which helped judges gauge public opinion and strike down the ban. The protests also inspired countless other small, bold acts of solidarity and defiance.

There's a time to breathe together, and there's a time to fight together. And it's never been more urgent to make sure we know which is which so that we can act effectively together.

If you have ...

Fifteen minutes: Find a local grassroots group or find a local chapter of a national organization mentioned in this book, such as Indivisible (www.indivisible.org), Our Revolution (https://ourrevolution.com), or 350.org, which has 182 chapters in North America (https://350.org). Get email alerts about the next local public action.

An afternoon or evening: Join a protest. Go to a public hearing on a key vote. Take a friend or child. See how it feels to join others and voice your opinion. A well-organized protest should leave you feeling empowered and connected. If it doesn't, find another group or a different action.

Righteous anger and passion: Help organize the protest you think we need. I've done it, and it can be both powerful and satisfying.

CULTIVATE FAMILY CONNECTIONS

37. Help Kids Feel Comfy

It's not easy being green.
— Kermit the Frog

"I am *not* biking to school," my seventeen-year-old daughter said with a finality I'd never heard before. For the previous three years of high school, Zannie had been on board with my planet- and health-saving program.

"Why not? You know I won't drive you a mile to school."

"My backpack is heavy, especially on lacrosse days, and it hurts my back."

"Let's get a bike rack for it," I suggested.

"Fine, but I'm *not* riding if you make me wear a helmet," she said.

The idea of her riding in rush-hour traffic with a bare head terrified me. I asked, "Does the helmet…mess up your hair?"

Her voice rose. "No! But I won't be the *only* senior riding to school in a helmet, especially when other seniors drive their parents' cars to school — right by me!"

"Ah," I said. "Okay, give me time to think."

In my teen parenting class, the teacher had recently said, "There are only two things a teenager truly fears: parental rejection and humiliation before peers." Zannie's reluctance to commute by bike landed squarely in the humiliation category.

I felt for her. It's not always easy to be the kid of climate activists. She'd been a good sport for years about our small house, front-lawn protest art, and the healthy, homemade (*boring!*) sack lunches she unwrapped in front of other high schoolers eating exciting, packaged foods or even takeout from nearby restaurants.

In describing the helmet impasse to my neighbor, Jen, I realized Zannie was wearing my old helmet. Jen zeroed in on that and pulled up photos of bright helmets at a nearby store — polka dot, floral, and leopard print. Grinning triumphantly, she said, "Let her choose one of these totally cute ones!"

Zannie and I made it through the bike-shop doors right before they closed, and the next morning, and every morning until graduation, she biked to school with a sleek black rack and gray (non-attention-getting) helmet.

The point of climate revolution is to live, think, and behave differently. But especially for tween and teens, *different* can feel like social suicide. In some situations and places, different can truly be dangerous.

To wage climate revolution while also empowering children and staying connected with them, it helps to first listen, understand what's behind their angst, and validate their feelings. Then, try to find climate-friendly solutions that honor those totally normal urges to fit in.

Some days, we will fail to find a solution. Sometimes, we may choose the plastic-and-polystyrene-wrapped takeout, just to preserve family harmony — and that's okay, so long as we don't give up the long-term climate fight. Because, though I battle full-time for the Earth and justice out of my love for both, what really keeps me going is my far-fiercer love for my children.

If you have...

Kids of any age: Read Kim John Payne's wonderful books: *Simplicity Parenting: Using the Extraordinary Power of Less to Raise Calmer, Happier, and More Secure Kids* and *The Soul of Discipline: The Simplicity Parenting Approach to Warm, Firm, and Calm Guidance — From Toddlers to Teens.* Also visit Payne's website (www.simplicityparenting.com).

Read Bruce Feiler's book *The Secrets of Happy Families: Improve Your Mornings, Tell Your Family History, Fight Smarter, Go Out and Play, and Much More.*

Read *How to Talk So Kids Will Listen & Listen So Kids Will Talk* by Adele Faber and Elaine Mazlish.

Join a support group, babysitting co-op, school parenting workshop, or parenting class. Ask for and offer help. Also, read the wonderful parenting essays in the magazine *Brain, Child* (www.brainchildmag.com) and on the blog Motherwell (https://motherwellmag.com).

38. Dig, Dawdle, and Dog

*Keep close to Nature's heart ... and break clear away, once in a while, and climb
a mountain or spend a week in the woods. Wash your spirit clean.*

— JOHN MUIR

Feeling low? Stressed? Disconnected from your kids, mate, or self?
This chapter is all about recovering our mojo and reconnecting to one
reason we fight for nature in the first place: It makes us happy.

If You Have ...

A nature deficiency: Our health and moods soar in the great outdoors, so
get everybody out in all weather and times of day:

Walk outside. Inhale deeply. Take in horizons, trees, birdsong. Ditch
shoes (even in the rain). Howl at the moon. Build snowpeople. Climb a
tree. Camp in the backyard (or on your balcony).

A dog: Dogs are a natural antidepressant, as are all pets. However, dogs
don't just offer unconditional love — they get you walking outdoors
every day in order to heed nature's call and get their exercise. So walk
the dog.

Restless kids: For a host of ideas and resources, read *Vitamin N: The Essential Guide to a Nature-Rich Life* by Richard Louv, which includes resources for meeting like-minded families. Then, take kids to the park or
send them into the backyard, while offering plenty of supplies for outdoor
play, such as the following:

sidewalk chalk	badminton
jump ropes	basketball and a hoop
coffee-can stilts	Frisbees
shovels, buckets, and a hose	hacky sacks
balls, all shapes and sizes	slacklines
bicycles	

For older teens, consider adding darts or bows and arrows (with
cardboard boxes or hay bales for targets).

The blues: Laughing is so good for our health that people even pay for laughing classes. Not parents — we have live-in comedians! However, sometimes kids need encouragement, so try one of these games (I've used them all successfully):

> Laughing Party: This is catnip for many kids between ages three and six. Find an excuse to laugh, and play it up, dramatizing hoots and snorts. Repeat until everyone gasps for breath.
>
> Toilet Paper Boots: Little feet can wriggle all the way through twelve-packs of toilet paper and stomp around. Make a hole in the plastic and slip feet between the middle rolls; the toilet paper package acts like huge leg warmers. Add dance music.
>
> Wear Your Whole Closet: Have kids put on five to seven layers of clothing and try to perform basic dance or gymnastic moves. Challenge them to run a simple obstacle course.
>
> King of the Bed: A classic. Contestants battle to push one another off of their parents' bed.
>
> Dish Towel Tag: The goal: Snap teens with a cloth napkin or dish towel without actually hurting them. This breaks homework tension and can quickly become a fun, house-wide chase.

Parental burnout: Please, let me nag you about self-care.

> Sleep more: A third of Americans are sleep-deprived. Fix that by turning off screens an hour before bed. Exile televisions, computers, and phones from the bedroom. Follow a regular sleep schedule. Get at least seven hours of sleep a night. As one well-rested mom gushed, "Sleep is God."
>
> Exercise aerobically and regularly: As you age, add in stretching, weights, and weight-bearing exercise (for healthy bones).
>
> Get vitamin D: Take supplements, get one of those full-spectrum therapy lights, and run outside when it's sunny (baring that skin!).
>
> Pray or meditate: Do this alone or, better, with others (people with spiritual communities report feeling happier than those without).
>
> Breathe deeply: Practice taking three deep breaths whenever you feel anxious, or in the middle of a conflict, to immediately relax. Teach your kids to do this.

Practice mindfulness: Try Bob Doppelt's Transformational Resilience resources, which apply mindfulness techniques to climate crisis anxiety (www.theresourceinnovationgroup.org).

Eat well: Eat whole foods and reduced processed ones. Ditch (unhealthy) white sugar and substitute honey, maple syrup, stevia, juice/fruits, rice syrup, or barley malt.

Make love: It's not just good fun, but actually good for your health!

Eat dirt: Really. Researchers are confirming what dirt-eating babies already know: Dirt contains bacteria that may lift moods and improve immunity. So, get dirty — and don't overscrub your garden carrots.

39. Bring on the Awe

Awe (noun): an overwhelming feeling of reverence, admiration, fear, etc., produced by that which is grand, sublime, extremely powerful, or the like.
— DICTIONARY.COM

In 1985, British climber Joe Simpson found himself injured and trapped alone in a crevasse in the Peruvian Andes. In his book *Touching the Void*, he describes slipping into a dangerous hypothermic state that shifted only when the sun finally rose.

"A pillar of gold light beamed diagonally from a small hole in the roof, spraying bright reflections off of the far wall of the crevasse," he writes. "There was a feeling of sacredness about the chamber, with its magnificent vaulted crystal ceiling, its gleaming walls..."

Mesmerized, Simpson was suddenly energized and determined to reach that sunbeam. He climbed from the crevasse and, over the next three days, crawled around countless other crevasses to arrive at base camp, hallucinating and near death. His endurance and luck were key, but awe is what roused him during that move-or-die moment. Awe can open our minds, first to our connection with the world and then, because we're paying attention, to new ways out of predicaments.

And awe makes us nicer. In 2015, researchers tested the behaviors of participants who had just experienced a sense of awe, such as looking upward in a eucalyptus grove for one minute. After experiencing — or even remembering — moments of awe, people acted more ethically, kindly, and generously. Even just watching magnified water droplets focused participants' attention on the world's beauty and complexity and activated a sense of being small in relation to a grand universe.

I don't know about you, but I find this all very heartening. What a great thing that when we humans let awe melt our hearts — and our sense of separation — we're more inclined to share our last cookie, open doors for the elderly, and act ethically toward others. That's lovely for its own sake, but isn't it wonderful that prioritizing humanity's experience of awe could conceivably help us escape the planetary equivalent of Simpson's icy crevasse? What if, before any vote to gut health care or roll back laws protecting our kids from toxins, our leaders were required to spend one minute experiencing awe?

I'm putting that on my wish list of laws to pass. In the meantime, here are ways to cultivate awe in our own families.

If you want to feel awe...

Tonight: Over dinner or at bedtime, invite everyone to share one moment that provoked wonder or awe. Or look at coffee-table nature books together. Sometimes I grab a few from the library and sit next to my teen for a five-minute awe break.

This week: Try some kitchen awe when making salad dressing: Fill a glass jar partway with olive oil. Let kids pour in balsamic vinegar. Hold it up to the sunlight. It's as entertaining as a lava lamp (but better because you can eat it).

Anytime: The next time there's a lovely sunset, full moon, lightning storm, rainbow, or if you're lucky enough to have them, northern lights, prod everyone outside to watch — or at least *look* through the living room /train/car/airplane window for ninety seconds.

When traveling with children: Make sure to include, when planning visits to museums, amusement parks, or relatives, ample time to explore nature wherever you are. We have found that the key to accessing awe is for people to feel unhurried; they become relaxed and open to what's around them. We've had some awe-filled moments even at highway rest stops — when we parents weren't in a rush to get back on the road and took time to poke around.

Trust that if you mine the world for beauty and share it with children, even cynical teens will eventually imitate you and seek out moments of awe on their own.

40. Exude Gratitude

My fourth-grader announced, as I sautéed rice and tofu for the lunch thermos, "That tofu thing doesn't taste good."

"What? It's your favorite!"

"It's gross lately."

"I'm making it as I always have. What's different?"

"I don't know," my child said, "but it's bad."

"Too much tamari? Not enough?" I heard myself patiently offering various remedies. As each was stonewalled, I grew annoyed.

Biting back a comment about the one-in-five hungry children in our own town who'd love that "gross" lunch, I stashed the thermos for myself and gave my child a PB&J and a kiss.

"Starting tomorrow, you'll make your own lunch," I said.

How did I raise an entitled brat? I wondered. I don't know what the future holds, but I want my kids to cultivate the "attitude of gratitude" that's key to a good life and loving relationships.

As I meditated on the idea of gratitude, though, I realized that I didn't model it very well myself. I often cataloged everything wrong with the world, but not enough about what was right. Steeped too often in this negativity, it was no wonder my child disdained my offerings. Our whole culture mirrors this sense of entitlement; we plunder the soil, water, air, and wildlife without regard for what's being sacrificed. If Earth really is a Mother, she must feel as I did that morning — pissed at being taken for granted.

I want my kids not only to appreciate what's given but to live with a sense of abundance and open-hearted gratitude. Counting one's blessings helps anyone feel better, but it's especially important for growing children. Studies affirm that young people who are grateful are happier, healthier, more positively engaged in community activities and school, and less materialistic and depressed. I also find that grateful kids tend to appreciate the Earth's beauty and want to protect it.

I resolved that day to live more gratefully. I listed my riches: full fridge, cozy home, close friends, parents still living, an easygoing husband, a self-healing body, rights to assembly. At dinner that night, I expressed gratitude for bountiful foods, miraculous water coming from our skies — and showerheads — and for the people I'm honored to call family. My science-minded husband chimed in with, "Yeah, it's cool that we're mobile, self-aware matter." My kids smiled, and when I asked them if they wanted to add anything, one said, "Good food," and the other said, "Friends and family."

Two weeks later, after many kid-made lunches, my child was thrilled when I agreed to make lunches again on a "see how it goes" basis. The first morning, eating cereal and toast while watching me assemble PB&J, apple, carrot, and cookie, my child said: "Have I told you lately how much I appreciate all the food you always make me?"

If you want to . . .

Encourage gratitude and appreciation: Engage children in meal-making to help them appreciate what's involved. Borrow children's cookbooks from the library. Leave them around for kid inspiration.

One night a week and on holidays and birthdays, ask everyone to say something they appreciate from the week and the year. Or institute a regular meal blessing. For inspiration, read *A Grateful Heart*, edited by M. J. Ryan, with prayers from around the world.

Help others: As a family, serve holiday meals to the homeless at a local soup kitchen. With kids, drop off everyone's nice but outgrown clothes at family shelters. Give cash donations on Mother's or Father's Day in honor of parents.

Help your own children donate to nonprofits they admire. Many children especially connect with Heifer International (www.heifer.org), which provides goats, chickens, or sheep for families in need worldwide.

Throw a reverse birthday party. Insist on "No gifts, just your presence." Feed your guests and gift them with things you no longer need — and compliments.

41. Time Travel

*"If only I had some grease I could fix some kind of a light," Ma considered.
"We didn't lack for light when I was a girl before this newfangled kerosene was
ever heard of."*

*"That's so," said Pa. "These times are too progressive. Everything has
changed too fast. Railroads and telegraph and kerosene and coal stoves —
they're good things to have, but the trouble is, folks get to depend on 'em."*

— LAURA INGALLS WILDER, *The Long Winter*

If you were to ask my daughter, Zannie, "What was the name of that girl
on the boat swing?" — referring to the toddler she met at a Renaissance
festival eighteen years ago — she can tell you. That's because Brianna
shared what was unequivocally the highlight of Zannie's first three years
on Earth.

The boat swing was a giant handmade wooden contraption with a
pirate feel, and the two girls, who met for the first time as they clambered
in — no adults allowed — handed over their tickets as the parents hov-
ered. A man with unnaturally large biceps asked their names, then sent
the girls flying. For the next few minutes, the toddlers squealed as they
sailed back and forth over our heads. They stumbled off the boat like
drunkards, bonded for life, and begged to go again.

Renaissance festivals celebrate life as it was lived long before our de-
pendence on fossil fuels. Visit one, and instead of playing video games
in arcades, your kids will watch live jousting matches. Instead of riding
merry-go-rounds with canned music, they will ride flesh-and-blood ponies as madrigal singers croon. Instead of Mickey Mouse and SpongeBob,
kids interact with festival-goers dressed as court maidens, blacksmiths,
and knights. They pet rabbits and goats and eat roasted turkey legs. It's a
great way to time travel.

Since then, we've enjoyed other festivals, Waldorf school fairs, off-
the-grid farms, living museums, and pioneer homesteads where my children have sheared sheep, threshed wheat, baked bread, churned butter,
ground cacao for hot chocolate, milked cows, picked blueberries, and
dipped dozens of beeswax candles.

These experiences help put central heating, mall shopping, and frozen pizzas into perspective for kids. They get a taste of how humanity once lived — and how 1.1 billion people worldwide still live — without electricity. Kids learn this is not only possible but it can be fun and empowering.

That's not to romanticize life without running water, electric lights, or flush toilets — obviously, those modern amenities vastly improve our lives and health. But for my own children, experiencing a human- or animal-powered lifestyle, even for an afternoon, expanded their appreciation for conveniences often taken for granted. It helped them understand that their lifestyle is actually an anomaly in the grand sweep of the human experience.

If you have...

$5 to $25: Camp. Leave electronics behind. Most kids love the break from modern life — and appreciate it when they return. (Irony alert: Wilderness gear can be both high-tech and pricey.) Many teens, once they've taken a wilderness survival class, prefer minimalist camping.

Take a field trip to an off-the-grid farm. Try the Laura Ingalls Wilder museums (http://littlehouseontheprairie.com/historic-locations-and-museum-sites) or the Association for Living History, Farm, and Agricultural Museums (www.alhfam.org).

$25 to $75: Attend a Waldorf school winter holiday or Mayfair, where children make wreaths and window stars, dye silk scarves, dip candles, or play dreidel games and Cake Walk. Find your nearby school at https://waldorfeducation.org.

Visit an Amish village that educates visitors about the centuries-old "Plain" lifestyle that relies on windmills for electricity and horse-and-buggies for transportation. Though often touristy, these villages can show children how simple living derived from deeply held religious beliefs can result, unintentionally, in a small carbon footprint.

Attend a Renaissance festival (do an online search for local listings). Try archery, javelin throws, hair braiding, ziplines, maypole dances, and if you're lucky, the pedal-powered snow cones my kids get every year at the Oregon Country Fair — free if they take a shift cycling.

RAISE
EMPOWERED KIDS

42. Throw Out the Baby-Safety Catalog

Fear is lucrative. Fear is big business.
— Mikael Colville-Andersen

One day when I was a new parent, a glossy catalog arrived with some vague "baby safety" name. It was thick and terrifying.

"Let's see what we're missing!" I suggested to my infant. Inside were countless vinyl, foam, and battery-operated products that would allow me to ignore my daughter for long stretches. I could video-monitor her from another room, leash her in public, and not only carry her in a portable playpen but, as I once did with gerbils, attach a net lid on it without asphyxiating her. At bedtime, I could program a stuffed teddy bear to sing her to sleep. Page after page after glorious, horrendous page promised baby safety and parental liberty.

At the time, we were far from gear- or fear-free. Our home contained countless ingenious devices to make washing, transporting, feeding, and rocking our baby to sleep easy and safe. Our living room proudly displayed that doorway jumpy thing. Art and I deployed outlet covers, toilet locks, a wood-stove barrier, and a monitor that piped our baby's waking cries into the kitchen.

As most new parents do, I feared that my child might suddenly stop breathing. I feared sudden infant death syndrome (SIDS), choking, drowning, and baby-nappers. But not enough to buy the expensive video monitor or the foam helmet and kneepad set for crawlers. Today, you can purchase $250 owl socks that monitor your baby's heart rate and oxygen levels and set off an alarm if something "isn't right."

I tossed the catalog. When our baby learned to sit by herself and play at our feet, we threw a pillow behind her. When she toppled, she sometimes hit the wood floor. She still got into college.

All parents must choose, through trial and error, products that fit their needs, budgets, and philosophy. High-tech gadgets can certainly save lives and sometimes be necessary (such as for babies with heart conditions, and so on). But for most families with healthy children, such products only stoke fears. Which is the point. It's an age-old formula: Activate fear, introduce life-saving product, get rich.

But I have competing fears. I fear filling our homes and children's bodies with off-gassing toxins. I fear exposing pregnant mothers to birth-defect-causing chemicals as they labor in factories to produce my baby's safety products. I fear the greenhouse gases being added to our already-choked atmosphere as these countless gadgets are produced.

The irony is painful: To keep children safe, we help ruin their world.

Moreover, aren't young children disempowered by all this strapping in, locking, surveilling, leashing, and restraining? We should absolutely keep children from toppling off balconies or petting alligators, but children also learn important skills, judgment, and agency through their own trial-and-error explorations of the world.

Life is unpredictable and I want my children safe. But personally, I try to let go of factors beyond my control — and learn to live with that supreme discomfort every parent feels sometimes — rather than try to eliminate every possible risk using all the plastic- and toxins-based materials at my disposal, regardless of their cost.

Everyone deserves safety, but that can be defined globally as well as personally. And one way to increase safety for all is with one small act of resistance: Reject the fear-mongering. Recycle the catalog. For safety advice, consult trusted experts, other parents, and most importantly, your own good instincts.

If you have...

Two minutes: Call the baby-safety catalog retailer's toll-free number to remove yourself from their mailing list.

Ninety minutes: Make an appointment with your health care provider and discuss fears and safety strategies. SIDS, for example, is now considered largely preventable.

An afternoon: Babyproof your home. Learn how to cut and serve food to avoid choking hazards for infants. Mothering.com has helpful parent forums on this.

Attend a local or online parenting class to learn more about which safety precautions make sense for your family.

43. Teach a New Set of Life Skills

I raised two self-centered, materialistic, screen-addicted kids. I wish I had a do-over.

— A dad acquaintance of mine

When my daughter was applying to college, I began worrying about everything I'd failed to teach her. She could manage money and school bureaucracies, but could she cook? Treat a first-degree burn? Talk a depressed friend off of a ledge? My sister-in-law, Laura, sent me her list entitled "Tracy's Badges." A badge was earned when every task on the list within each of eight categories was accomplished. Neither Tracy nor her brother, Brian, were allowed to leave home until they had earned badges proving their mastery of the basics of managing kitchens, wardrobes, homes, finances, cars, relationships, and first aid.

Had I started earlier, Zannie might have earned her eight badges, but she only changed one tire, sewed on one button, and prepared one stir-fry before she called it good and left for college badge-less. But the "Tracy's Badges" list inspired me to make a new list that includes not only basic skills children need in today's world but those that empower them to build a fossil-free world, live generously, practice self-care, and thrive in today's world, no matter how fierce its storms.

To create this list, I studied popular online lists of skills children should possess by certain ages, as well as one by Julie Lythcott-Haims, author of *How to Raise an Adult*, whose concern is the readiness of young adults about to leave home. I left out some obvious skills, such as time and homework management, and I left the housecleaning chores vague, though those are important, to make room for underappreciated yet necessary skills. Finally, ages are approximate, and every family is unique, so tweak this list so it fits your values and lifestyle.

Help children do the following, at age ...

Five and under:

Comfortably hold bugs; gently catch (and release) frogs and fireflies. Express gratitude for food, toys, and gifts.

Make homemade cards and thank-you notes.

Choose clothes and get dressed; brush teeth and wash face.

Help with household tasks, including cooking, cleaning, laundry, and pet and plant care.

Begin mastery of shoe tying, coat zipping, buttoning, bike riding, swimming, skating.

Learn own name, parents' full names, address, and phone number.

Know at least six short songs by heart.

Six to nine:

Safely ride bicycle in town, and load bike onto city bus.

Build a fire and roast tofu or chicken hot dogs over flames, or potatoes in the coals.

Learn more homemaking skills, simple meal prep, knife use, and proper composting.

Make homemade gifts (such as candles, simple jewelry, artwork).

Make simple household cleaners from vinegar, baking soda, and water.

Assess whether clothes truly need washing/machine drying.

Count money, give change. Manage allowance. Save for longer-term goals.

Know how to get help in an emergency (neighbors to contact, calling 911).

Learn to listen to and comfort others who are sad, angry, or hurt.

Play an instrument (even a kazoo, recorder, or bongo drum) and/or sing a lot; dance regularly.

Meet with or write letters to school principal, mayor, or representatives about key issues.

Ten to fourteen:

Safely handle kitchen knives, pocketknives, matches, and lighters. Know several rope knots.

Plan and prepare a meal with protein and veggies; broil or bake foods.

Read food and cosmetic labels; understand which ingredients harm human, animal, or planetary health.

Use basic hand tools: hammer a nail, drive a screw, use wire cutters and pliers.

Take a babysitting course and/or lifeguard class; learn basic first aid and CPR.

Start getting work experience (paid or volunteer); keep saving money.

Be comfortable stating or negotiating fair pay rate for jobs, such as babysitting or pet care.

Compare quality and prices when shopping; be able to calculate unit prices.

Oil bicycle chain and fill bike tire.

Take spiders outside comfortably.

Detect when any kind of food is spoiled.

Make some foods from scratch (such as muffins, cookies, granola, soup).

Understand uses of medicine and the dangers of misuse or overuse.

Make simple projects (like napkins or cloth gift bags) on a sewing machine.

Testify at city council or public hearing about a local issue they care about.

Know the signs of depression, anxiety, eating disorders, and other mental health risks.

Make good eye contact with adults during conversations.

Take a class in reproduction, sexual education, and self-care.

Get daily exercise inside or outdoors.

Type proficiently; be able to search internet without distraction from click-bait.

Be able to play three songs on a ukulele, guitar, or piano while others sing along.

Fifteen to seventeen:

Clean refrigerator, defrost freezer, change vacuum cleaner bag, clean stove, unclog drains.

Change flat bike tire, perform basic maintenance, install fenders and lights.

Charge an electric car or fill car with gas; perform basic maintenance.

Develop zero tolerance for texting or drinking while driving.

Clean a cut or scrape and bandage it, treat a burn, wrap a sprain.

Safely light a gas stove or oven or outdoor grill with a match.

Hang a framed picture.

Manage money: understand fraud protection and balance checkbook.

Buy and bargain for used items.

Find credible information sources; critique news and develop media literacy.

Set healthy limits with screen use, including ample nightly brain rest (phones off nine-plus hours).

Prepare a resume; interview for and get a job.

Volunteer or regularly do nice things for others, such as help unload elderly neighbors' groceries.

Reset an electrical breaker, turn off water main, stop a running toilet, unstop clogged toilet.

Learn about healthy sexuality, and about what consent looks and feels like.

Memorize one song from another culture.

Eighteen and up:

Earn money, manage running of a household, manage personal finances.

Manage school and work deadlines; find healthy balance of work life and social life.

Develop good self-care habits; use exercise, sleep, and mindfulness tools to avoid stress.

Safely conduct Craigslist (or eBay, 5miles, Poshmark, or Kiiboo) transactions, and rideshares and car hires.

Understand apartment lease and other contracts, such as for rentals like Zipcars.

Understand interest rates and the serious implications of consumer debt or student debt.

Comfortably speak with doctors, professors, bike mechanics, landlords, clerks, farmers.

Navigate a campus or a new town.

Take risks, handle personal problems with compassionate communication skills.

Know our constitutional rights; know what to do in case of police encounter or arrest.

Register to vote, learn about social issues from reputable sources. Vote.

Memorize one love poem.

44. Meet Some Badass Kids

We cannot trust that the adults alone will save our future. We have to take our future in our own hands.
— FELIX FINKBEINER

O ne day, one of my nephews, after only four years on the planet, had received enough petrifying information about his prospects that he announced, "Mom, I don't want to be alive anymore."

Alarmed, his mom pulled him onto her lap and asked, "Why?"

"The Earth will die, and I don't want to be here when everything's dead."

Children absorb a steady stream of disturbing stories about drowning polar bears, deadly superstorms, and dissolving starfish, but rarely do they hear about the countless kids worldwide who feel exactly as they do — and who are rolling up their sleeves to save the planet.

Take Xiuhtezcatl (pronounced "Shu-tez-cot") Martinez, a seventeen-year-old from Colorado. He is youth director of Earth Guardians, an international "youth-led group of mobilized badasses," as they call themselves. He and Earth Guardian crews from Cape Town to Istanbul and Bogota organize rallies, leadership trainings, tree-planting parties, school walkouts, and more to "disrupt the global political leadership status quo." Xiuhtezcatl is also suing his governor and the president over destruction of the climate (See "Sue the Grown-Ups," page 244). And because he's a social media wizard, Earth Guardians reaches 1.5 million kids worldwide.

That's a lot of kids. More join daily. Yours can, too.

Xiuhtezcatl is also an indigenous hip-hop artist who, with his brother, Itzcuauhtli ("Eet-squat-lee"), gets audiences up on their feet, stomping and singing to "Be the Change!" When the brothers rocked a 2015 environmental law conference, my son's friend Wesley became so energized by the workshops and the empowering music from kids his age that he told his mom, "I'm going to become an activist."

That year, as a tenth-grader, Wesley cofounded, with Corina MacWilliams, a combined Earth Guardians and 350.org club, beginning with seven students at a backyard pizza party. A month later, forty-five high schoolers signed up at the student club fair, and "EG/350" was launched. Since then, they've relentlessly led rallies, pushed regional climate recovery legislation, created protest art, challenged elected officials, and boisterously supported plaintiffs in the youth climate lawsuit. The morning after Trump's election, they organized a student walkout. More than four hundred students — and the media — joined them.

I've known several of these students since they brandished baby teeth, and I can see how empowering these experiences have been for them. Some of them are now discovering a side benefit: College admission and scholarship committees love youth climate justice leaders. Just sayin'.

If you have...

Ten minutes: Check out Earth Guardians (www.earthguardians.org), whose mission is to "grow a resilient movement with youth at the forefront by empowering them as leaders and amplifying their impact." Get newsletters, join a kid lawsuit, or invite kids to explore what badasses are doing on the other side of planet Earth.

Get inspired by Indian Youth Climate Network's website (http://iycn .in). With kids, watch short films or a slideshow about how Indian youth — who make up more than half of India's population and know they're vulnerable to severe climate impacts — are organizing nationwide for climate action.

Also visit iMatter (www.imatteryouth.org). Help kids give their town a Youth Climate Report Card.

One to two hours a month: Support new climate-action clubs or a youth wing of your local group. Help kids with logistics, activities, or snacks for meetings (a key ingredient, my kids say) as they grow their leadership abilities. Post photos and stories from their events and gatherings.

Help kids hear from other young activists. Invite youth leaders to classes, scout clubs, or church youth groups. Can't find local activists? Sponsor an Earth Guardian from a nearby city to mobilize your group.

Time for a book: Read *We Rise*, a guidebook for climate action, by seventeen-year-old Earth Guardians youth director Xiuhtezcatl Martinez.

45. Let Kids Make Laws

We spoke up to city council, and we were like, "We really need to get this passed. We really, really want this and need this to get passed," and it happened, and I'm really happy!
— Avery McRae, YouCAN member, age ten

It was the Big Night for climate. Our family calendar had stars and smiley faces all over that Monday, so we headed downtown to grab seats at city hall. My kids had never attended a city council meeting, but I said, "The councilors vote tonight on the first-ever science-based climate-recovery law. You're the face of the future they need to see while deciding."

Bay and Tayo, two brothers in grade school, had asked their representative to sponsor the nation's first climate-recovery ordinance that would hold our city accountable for cutting carbon emissions at the rate science, not politics, says is necessary — to return atmospheric concentrations of CO_2 to a stable level of 350 parts per million by 2100. Bay and Tayo's mom, environmental lawyer Julia Olson, understood that science-based greenhouse gas regulation, stalled by political gridlock on the federal level, could grow from the local level.

Julia had helped her sons start the group Youth Climate Action Now (YouCAN), and for ten months its dozens of members had lobbied Eugene's mayor, city council, and public. YouCAN families wrote letters to the editor to support the unprecedented law that would bind the city to carbon-neutral operations by 2020, a citywide carbon budget, and 50 percent less community fossil-fuel consumption by 2030. This works better with local officials than with federal ones, which is why YouCAN's efforts champion a decentralized approach. We can easily meet our mayors, the logic goes, and affect their policies far more effectively than a president's. If citizens in every city do this, we can lead from the bottom up.

With help from a local artist, the kids had designed and painted a mural on a bike shop facing a busy street. At the unveiling of the mural — which features joyful children, a polar bear, and a giant bicycle with a snowy blue-green Earth as its front wheel — the children proudly gave media interviews and called on the public to support the science-based climate recovery ordinance.

On voting night — a hot night in July 2014 — hundreds flooded city hall, mostly in support of the children who, one after the other, pleaded with councilors to pass the law. Then the mayor finally asked the nine councilors, "All in favor?"

Seven hands rose. Kids and parents silently high-fived, hugged, and slipped outside to freely hoot and happy-dance while reporters snapped photos.

My teens asked, "Wait, our city council really just passed the first climate-recovery ordinance in the whole country? And our being here actually helped?"

"Yes. And yes. First in *any* country, actually," I said.

They smiled, nodded, and said, "Cool."

YouCAN's campaign has moved to several other cities. Bay and Tayo, now experts, can help other kids to create this front wave of local laws. One day, they can tell their kids that they helped pass a science-based climate-recovery law — the first in the whole beautiful world.

If you have...

Thirty minutes: As a family, watch the inspiring fifteen-minute video "TRUST: The Climate Kids" (at Our Children's Trust, www.ourchildrens trust.org/trust-the-climate-kids), about how children passed this powerful new law. With your kids, brainstorm a mural in your town, identify a building to host it, and call the owner. Ask if you can paint a planet-loving mural on their wall.

One hour: Explore what it would take for a group of young citizens to pass a local climate-recovery ordinance where you live. For help, contact Our Children's Trust or download and read their free YouCAN how-to manual (https://www.ourchildrenstrust.org/learn-how-to-start-a-you-can-chapter).

Perseverance: Take your children to meet with your mayor or council representative to discuss what kind of science-based climate-recovery actions your city can take.

Host a potluck to discuss forming a children's group for local climate action — including passing a local climate-recovery law.

Organize a mural-painting gang, buy paint and supplies, and change the face of your community.

46. Skip the Kid Shock Treatment

Always look for the people who are helping, [my mother would] say. You'll
always find somebody who is trying to help.
— MISTER ROGERS

One evening, my husband, Art, and our ten-year-old daughter, Zannie, visited a neighbor recuperating from knee-replacement surgery. The man described his surgery in detail, not noticing her blanching, and when he got to the part when "they sawed my leg in half," Zannie tipped over in a dead faint and convulsed on the floor.

This is how many adults talk about global warming around kids — without censoring themselves. Or with hopelessness. Or not at all.

Good resources are hard to find. Do an online search for "How to talk to kids about global warming," and you mostly get classroom resources for conveying child-friendly science. These tend to avoid any mention of politics or of industry-funded junk science; instead, they serve up plentiful doses of "Take action by turning down the thermostat!" It's the climate equivalent of a doctor dropping a terminal cancer diagnosis, then sending the devastated patient off with vitamin C and a cheerful wave. Children need something in between that's neither a system shock nor out of touch with our climate reality. That's where parents come in. They often know when to deploy a take-it-easy approach for scary or confusing news while also offering positive ways for their kids to engage in solutions.

On the climate front, Art and I try for truth, emotional support, and empowerment without either overwhelming our kids or pretending everything is fine. When our kids were very young, we never shared scary climate news (or let TV bring news into our home; see "Unplug the Kids," page 145). When we drove past new, shocking forest clear-cuts and our children expressed sorrow, we empathized, saying, "Many people feel as you do and are working hard to stop timber companies from doing that, and they'd love your help."

Because it's true. And this message lets our children know that others feel that same grief and rage and channel it into action.

Pairing bad environmental news with emotional support is critical for children of any age because underneath their questions about vanishing forests, polar bears, and glaciers lurk deeper existential questions: *Am I going to be okay? Have you adults got this?*

Art's and my response has been to immerse our kids in empowering books, mountain hot springs, and the good fight *against* Stupid stuff and *for* Smart stuff. Now that our children have matured into young adults, we share our political opinions and planetary worries more freely and ask to hear theirs. Because they're increasingly exposed to the world's cynicism, it's especially important to remind teens that many people are working hard to make the world better.

If you have...

Young children (age seven and under): For Mister Rogers's tips on supporting children through tragedies and bad news, visit the "Tragic Events" page at the Fred Rogers Company website (http://www.fredrogers.org /parents/special-challenges/tragic-events.php).

Read picture books together that show heroes taking positive action in the face of an environmental challenge:

> *One Plastic Bag: Isatou Ceesay and the Recycling Women of the Gambia* by Miranda Paul, illustrated by Elizabeth Zunon
>
> *Ada's Violin: The Story of the Recycled Orchestra of Paraguay* by Susan Hood, illustrated by Sally Wern Comport
>
> *Follow the Moon Home: A Tale of One Idea, Twenty Kids, and a Hundred Sea Turtles* by Philippe Cousteau and Deborah Hopkinson, illustrated by Meilo So
>
> *Nights of the Pufflings* by Bruce McMillan

Middle graders (ages eight to twelve): See "Learn the Fossil Fuel Nitty-Gritties" (page 66), with its kid-friendly introduction to fossil fuels — and clean-energy solutions.

Teens: Suggest they read *The Legacy of Luna: The Story of a Tree, a Woman, and the Struggle to Save the Redwoods* by Julia Butterfly Hill.

47. Share Tales of Perseverance and Triumph

After nourishment, shelter, and companionship, stories are the thing we need most in the world.
— PHILIP PULLMAN

Nothing inspires me more in this long quest for climate justice than juicy stories about people with a lot of chutzpah. I feel so much better after tucking in with an exciting, real-life tale about people doing outlandish, world-changing things and having it all pretty much work out. Spending time with extraordinary people on the page or screen — from Susan B. Anthony to Winona LaDuke to young Ghanaian disability-rights activist Emmanuel Ofosu Yeboah — energizes my battle for climate justice, as well as my quest to become a better person.

So I've created this list of inspiring books and films for families, teens, and young kids (similar to the list of climate movies in "Host a Living Room Film Fest," page 168). These recommendations celebrate real-life heroes and their epic struggles to survive, realize impossible dreams, and/or bring justice to the world. They also share important lessons and even practical strategies that ordinary people (like us) can use to resist and triumph over injustice and brutality.

For more suggestions, talk to your local librarian and visit Common Sense Media (www.commonsense.org). Also, please consult book and movie reviews to ensure each selection is appropriate for your child.

Finally, consider declaring an Inspiration Night. Bake cookies, make special treats (like "Forrest's Too-Spicy-for-Mom Popcorn," page 170), and cozy in to read aloud or watch an uplifting film together.

If you want...

Good reads for young kids:

Dolores Huerta: A Hero to Migrant Workers by Sarah Warren, illustrated by Robert Casilla

Drum Dream Girl: How One Girl's Courage Changed Music by Margarita Engle, illustrated by Rafael López

Emmanuel's Dream: The True Story of Emmanuel Ofosu Yeboah by Laurie Ann Thompson, illustrated by Sean Qualls

Harvesting Hope: The Story of Cesar Chavez by Kathleen Krull, illustrated by Yuyi Morales

Rad American Women A–Z: Rebels, Trailblazers, and Visionaries Who Shaped Our History…and Our Future! by Kate Schatz, illustrated by Miriam Klein Stahl

Ruby Bridges Goes to School: My True Story by Ruby Bridges

Sit-In: How Four Friends Stood Up by Sitting Down by Andrea Davis Pinkney, illustrated by Brian Pinkney

Any book in the popular Who Was? early reader series, including books about: Harriet Tubman, Anne Frank, Nelson Mandela, Susan B. Anthony, Sitting Bull, Frederick Douglass, Rosa Parks, Martin Luther King Jr., and Jane Goodall

Good reads for teens (and adults):

Brown Girl Dreaming by Jacqueline Woodson

I Am Malala: The Girl Who Stood Up for Education and Was Shot by the Taliban by Malala Yousafzai (young reader edition available)

Long Walk to Freedom by Nelson Mandela

March by John Lewis and Andrew Aydin, illustrated by Nate Powell

My Beloved World by Sonia Sotomayor

The Spark: A Mother's Story of Nurturing, Genius, and Autism by Kristine Barnett

They Poured Fire on Us from the Sky: The True Story of Three Lost Boys from Sudan by Benjamin Ajak, Benson Deng, and Alephonsion Deng

Turning 15 on the Road to Freedom: My Story of the 1965 Selma Voting Rights March by Lynda Blackmon Lowery

PG-rated biopics for young viewers:

Gandhi

Hidden Figures

Norma Rae

Queen of Katwe

Rabbit-Proof Fence

Rudy

Sound of Music

Stand and Deliver
Temple Grandin

R-rated biopics for mature teens:

42
127 Hours
Billy Elliot
Erin Brockovich
Long Walk to Freedom (unrated)
Milk
Mni Wiconi: The Stand at Standing Rock (unrated)
My Left Foot
Pride
Selma
Touching the Void

48. Take Kids on a Real-Life Hero's Journey

Do or do not. There is no try.

— YODA

Throughout childhood, my son, Forrest, saved the world dozens of times each week. In kindergarten, he grabbed cape and wooden sword to battle backyard monsters. In elementary school, he drew epic battle scenes in which he and Batman vanquished the Joker. In middle school, he read countless books, then built new plots on top of whatever fictional world he inhabited that week.

It was honor and heroism with clarity of purpose: Slay the dragon. Destroy the Ring. Stop Joker from nuking Earth — all while keeping friends and family safe. But faced with a real-life planetary emergency, he, like many of his friends, can flounder. Doesn't his generation recognize how much we need them now? In past generations, young people were needed to help run the farm or family business. During World War II, kids collected tin scraps to help the victory effort.

Yet today, in an epic waste of societal resources, youthful heroism is channeled almost solely into virtual battles. Perhaps the statistics about the alarming — and rising — rates of youth anxiety, depression, suicide, and self-harm offer hints about how our children feel in their roles as passive consumers. Moreover, the preoccupations of social media, TV shows, and video games may be numbing kids to the profound crises in the natural world they need for survival.

What if our children confronted what actually threatens them? What if, instead of fictional Vaders, Voldemorts, or Jokers, they confronted global warming and its driving forces? Our climate recovery mission has every element of the classic hero's journey: an imperiled planet, a clear mission, dangerous madmen, wise elders, fine companions, and heaven knows, a ticking clock. The drama doesn't get any more classic than this (which is, of course, more fun in stories than real life).

But here's the cool part: Heroes grow in skill and wisdom *because* of the antagonist — as long as heroes get a little help. Heroes always get overwhelmed, and they look to friends for help and companionship, and to the wise elder for guidance. Luke has Obi-Wan and Yoda. Katniss has Haymitch, Harry has Dumbledore, Dorothy has Glinda, Odysseus has Athena. The list goes on.

Our kids have us, their parents. And we have our own advisors — counselors, teachers, experts, and elders — who help us to guide our children through this unprecedented eleventh hour. When we claim our roles, we can help them gain skills, understand their crucial roles, and just maybe, triumph.

If you want to...

Tell the story: Be the one to present the global-warming story to your kids. Shape it in a way that's appropriate to who they are — their age, maturity, and interests. This could be a "once upon a time" story for young kids or simply your take on the news your teens are already absorbing. Choose one event that represents the most important issues. Many are already named in this book: Recount the fight over the Dakota Access Pipeline (page 238). Describe renewable-energy champions such as wind (page 72). Discuss all those who spoke out in support of the Paris climate agreement (page 76). Name the villains — oil companies, climate-change deniers, President Trump, apathy — who undermine efforts to fight global warming. Let kids know they're living the world's most epic adventure story ever.

Read the story: For recommended books and movies, see "Share Tales of Perseverance and Triumph" (page 122) and "Host a Living Room Film Fest" (page 168). Also read *Heroes of the Environment: True Stories of People Who Are Helping to Protect Our Planet* by Harriet Rohmer or *12 Children Who Changed the World* by Kenya McCullum.

Be the story: Together, brainstorm one heroic action that your child might like to take. Being part of something important and larger than our individual selves is good for kids. Help them feel like true warriors in their own real-life heroic journey.

49. Re-story Our Future

I challenge my students to tell me about an "eco-topian" future, to get them thinking creatively: What kind of world do I want to live in?

— TIM INGALSBEE, sociology teacher

"I want to write a story about the Earth being destroyed," said my thirteen-year-old son, Forrest, over French toast one morning, years ago. He'd just finished reading James Dashner's dystopian novel *The Maze Runner*.

"How about writing a story about heroes who *save* the world by toppling corrupt governments that are destroying the climate?" I suggested.

"You always say that," he accused. I was both surprised to hear I'd suggested this plotline before and disappointed in his failure to recognize its potential.

"But it's true," I argued. "You'd break new cli-fi ground!"

He craved world destruction, though, so to encourage his literary aspirations, I helped him work up a plot for his next attempt at a YA thriller.

What is it with the smoldering-planet thing lately — especially for young audiences? Type "dystopia" into Common Sense Media and you get 611 options. "Utopia" only gets 46 (several of which, like *Dirty Dancing*, seem misfiled).

If the world weren't actually in danger of ending — irrevocably altered by climate destruction and in danger of becoming truly hellish — my teen's interest in the apocalypse wouldn't disturb me so much. Why not create uplifting climate-change fiction (aka cli-fi)? I'd buy it!

Parents can help kids "re-story" the future, meaning both "restore" the future and "tell a new story" about it. We can steer children — and ourselves — toward life-affirming stories of nature's healing power (like *The Secret Garden* by Frances Hodgson Burnett) or people protecting the Earth (like *Hoot* by Carl Hiaasen).

Discuss with older children how Earth-saving or Earth-destroying stories make them feel: hopeless or empowered to make change? Also keep in mind a 2013 New School for Social Research study that found reading literary novels can increase our empathy. Surely, readers of literary cli-fi would develop more compassion for people and animals suffering due to climate change, no?

This compelling information hasn't yet prompted my son to write cli-fi with a happy ending, but stay tuned. I'm still working on him.

If you have...

Fifteen minutes: With your kids, brainstorm inspirational plots for a good climate revolution story: Perhaps a family musical about nuns who stop a pipeline (like the Pennsylvania nuns who built a chapel in a pipeline's path). Or a romantic comedy about a petrol-geologist falling for a wind turbine technician. Or a thriller about indigenous peoples defeating a coal mine. Don't judge ideas, just let 'em rip!

Or watch laugh-out-loud fake commercials and deconstruct their messages and themes, like "Nature Rx" (warning, contains one bleeped f-word) and "Exxon Hates Your Children" (search these titles online).

A few hours: Challenge a group of kids to a writing contest: Who can write the best short story (say, five hundred words) set in the year 2050 about how humans ended climate destruction? You could make this more formal and announce it through your local school or climate group.

Or invite kids to make the best three-minute video commercial from the point of view of endangered species who are advertising all the ways humans will benefit by saving them. Encourage humor, and give kids permission to poke fun at adults — especially politicians and corporate executives, but really, all grown-ups — for messing up their world. Incite comedic rebellion. Make it fun and manageable. Then, reward the winners (or all participants) with small prizes. If possible, find or create a venue for sharing their masterpieces with others, such as through school publications, TV stations, or at a school presentation.

An urge to mentor young activists: Establish an annual scholarship for college-bound activists who write a winning essay on how their generation will create a fossil-free future. If you want, create the scholarship through a local nonprofit (like a local chapter of 350.org), offer a real prize (like $500), and showcase the winning entry in a publication of some kind.

50. Discover the Wonders of Bribery

Just a spoonful of sugar helps the medicine go down.
— Mary Poppins

I was stuck. The Bijou movie theater was showing *The Yes Men Are Revolting* for one last night. My fourteen-year-old, tucked in with comics, had zero interest in watching middle-aged guys play pranks on corporate climate polluters.

Of course, I had to go — creative activism, climate, and humor all rolled into one movie! — but I didn't want to leave Forrest alone for a third evening that week. I've always followed an "it's optional" policy on climate stuff — it's my thing, right? — so I wasn't going to force him to come.

Then I remembered our family's make-you-happy, make-me-happy household harmony strategy: "I need an orderly house. You need a waiver, a check, and a ride for soccer. Do your chores and we've got a deal." This beats haranguing every time. We often plan a movie night (teen preference) after game night (parent preference). One year, on a Grand Canyon trip, our hike-resistant middle schoolers racked up two dollars in gift shop credit for every flat mile walked and four dollars for every uphill one. Eager to "earn miles," they led, hunting for lizards while Art and I dallied. It was bliss — for less than the cost of the microbrews we didn't then need from the hotel bar.

I could similarly sweeten this Yes Men movie deal just because it mattered to me. He was free to say no.

"What bribe can I offer?" I asked.

His eyes lit up. He suggested the one thing I have always steadfastly refused to do. "Buy me candy at the movie theater." I struck a deal, and he immediately refined it, citing my oft-repeated "movie candy's a rip-off."

He asked, "Will you give me what you'd spend at the theater, but stop at the grocery store on the way? I'll get twice as much candy."

I feigned resistance but complimented his shrewdness. Armed with his double dose of Skittles, Forrest was soon chuckling at the Yes Men as they impersonated Halliburton oil executives rolling around a beach in "Climate SurvivaBall" suits. On the way home, we talked about his favorite parts and the power of humor.

I won't claim any magical connection between that film and my son's actions later, but I will say this: We had fun that night. He saw a new kind of climate action that was bold, funny, and effective. And the following fall, he was part of forming a youth climate club at his high school. In the years since, he has also chosen to show up for every major local climate march and Our Children's Trust's courthouse rally.

I hope he sees these as important, fun, and empowering, but it certainly never hurts that, afterward, I also sometimes treat him to takeout from his favorite local Thai restaurant.

If you have...

Five minutes: Identify one thing that brings you happiness but requires older-kid buy-in you rarely get, such as help with house chores. Ask your kids what makes them happy that you can do for them. Bargain.

Thirty to sixty minutes: At a family meeting, practice striking bargains. We've found this empowering for kids, who — if the discussion is loving, unrushed, and geared toward solutions, not finger-pointing — appreciate being consulted on how to make family life more harmonious. For ideas, read Bruce Feiler's great chapter on family meetings in *The Secrets of Happy Families*.

An afternoon: Apply bribery to climate action: Choose an event, such as an afternoon art build or rally, that could benefit from your child's participation and from which your child would learn and grow. Trade this for an afternoon at the pool — and throw in ice cream. For a big march, offer a night of paintball or minigolf.

On the day of the event, take extra care to ensure a fun, relaxed, and connected family experience. As my mother once said, "Set any important activity as you would a precious gemstone."

51. Model Love, Always

Darkness cannot drive out darkness; only light can do that. Hate cannot drive out hate; only love can do that.
— MARTIN LUTHER KING JR.

When President Donald Trump writes abusive tweets while on the toilet, I'm horrified. And yet, this is the perfect encapsulation of what's wrong with public discourse in our country today — the president of the United States leaves steaming piles in both the physical and the political world simultaneously. The only sane response is to flush them away.

Too bad it's not that easy. When public political discussions become caca-flinging fights, everybody sees it. Everyone gets dirty, including children, who are watching and learning. And if our elected representatives engage in this coarse, denigrating talk daily, it's no wonder that so many Facebook comment threads regularly devolve into brawls, as Facebook friends assail their coworkers, bosom buddies, and relatives over their opinions, however politely expressed.

That's why it's more important than ever for parents to model appropriate responses to nasty comments — no matter where they are encountered or what political views they represent.

Because I speak and write in public to mobilize people to climate revolution, I have had to devise a policy around confrontation: I loudly demand climate justice on stage, on the opinion page, and on signs I carry in the street. I assail the positions of my representatives who green-light Stupid, planet-killing projects. I protest billionaires like the Koch brothers when they wield their unprecedented power in ways that threaten the environment. But I don't attack them personally. I don't mock their spouses. I criticize behavior and policy, not the person. As my kids watch, I try to embody the dignified, nonviolent opposition to injustice epitomized by leaders such as Gandhi and Martin Luther King Jr. In the 1980s, when I was arrested during civil disobedience protests over the covert US war in Central America, I acted peacefully toward police officers. Being dignified, however, does not mean being quiet or obedient (see also "Break Human Laws (to Obey Nature's Laws)," page 243).

In all my interactions, whether in a speech to a crowd or in an email to friends, I try to not rant. Instead, I try to issue invitations — "Read this, attend this, take this action" — that leave people free to accept or decline without guilt. Guilt-tripping only inspires defensiveness and hostility.

But most of all, I avoid personal disagreements and conflicts online and in social media. Written words — especially when they are hastily composed and dashed off instantly for everyone to see — are too easily misunderstood, misinterpreted, and perceived as hurtful. Talking to someone in person is different. One on one, we have countless nonverbal cues that convey our feelings and intentions. How different is it to ask someone in person — with warmth in your voice, a twinkle in your eye, and glasses of lemonade between you — "Are you open to hearing how I feel when you say [insert upsetting comment here]?" than to write those same words in an email?

When people post anonymous comments disagreeing with an online article I've written, my policy is to not reply. This both sidesteps any nasty showdowns and conserves my precious time. Any response I have to whatever is raised is better crafted into a new article and offered to a bigger audience.

If someone I know disagrees with me on Facebook, I never duel. I usually "like" their comment simply to acknowledge them, and if a friend persists, I laugh it off: "Ha-ha, we'll never agree on that, right? How's your puppy?" I can't recall a single "Amazing Grace" moment when a Facebook friend changed a political opinion because of my pithy repartee or labored explanation.

Last, I don't process strong feelings about anything with colleagues or loved ones — particularly anger or hurt — by text and email, unless I know that a conversation isn't possible and not writing would worsen the problem. That's how I want my kids to behave, so that's what I try to model.

One big reason in-person conversations, particularly with lots of people, usually go more smoothly than electronic ones is because of our miraculous mirror neurons, which make us smile, cry, or tense up when we see others smile, cry, or tense up. This can't happen over email or text. When we talk with another person, mirror neurons move our muscles without our knowledge, working nonstop to harmonize human emotional states. This helps foster empathy, which leads to compassion, and they're why love and fear are contagious. In any situation, when one person boldly leads with love and understanding, it encourages others to do the same.

So, as parents who know their kids are watching, the single-most-important thing we can do is throw love bombs everywhere. Post invitations to your family's make-the-world-beautiful adventures, then post photos of the fun you're having, the cool people you're meeting, and how you're growing together. Cheer triumphs. Forgive others — and yourself — for mistakes. Appreciate everything and everyone.

Because, in the end, lovefests aren't just way more fun than rants. They're way more effective. Tweeting fear is contagious, yes, but so is expressing kindness, gratitude, and happiness. And love. People like to be around happy, loving people and are much more open to ideas, such as, say, joining a climate revolution, if they know the journey will be fun and the community kind. This means that one of the greatest gifts we can offer our kids is to model how to use love's unparalleled power to joyously build the future we all want to live in.

If you or your child feel...

Angry about a political issue: Respond with solutions, not rants or lectures. Take positive actions and invite people to join you. Be fun to be around. Celebrate good policies and the people who work hard to make them. Shower attention on the countless good people in public office and government valiantly trying to make things better. And always make sure that you voice your protests to those who are truly responsible. For instance, when music programs are cut, it's not necessarily the school board's fault, but that of lawmakers who cut education funding.

Bullied online: Ignore trolls and strangers who bully. If a genuine friend or relative's comments require a response, address them in person (if possible) and using nonviolent communication (see "Tame Your Tongue," page 173).

Inspired to advocate: Beware of "clicktivism" (sharing online articles and petitions). It's important to shift public perceptions, but nothing replaces real-world activism. Calling, showing up, and organizing together for real-world, achievable goals will change our system precisely because it takes time. When constituents sacrifice their time for a political issue, representatives understand they truly care.

Hopeless: Take action, even a tiny one, with another person. Perform it with love and gratitude.

52. Teach Civics — Not Reading — to Kindergartners

Every child should learn arithmetic — and how to testify to legislators.
— MARY C. WOOD

I remember exactly one lesson from kindergarten: Don't keep guppies in a shallow dish meant for turtles or you'll find them, as we did, stuck to the bottoms of our shoes. The rest of kindergarten was a happy blur of jump rope and graham crackers. The goal was to get us used to school. No alphabets adorned the walls, but I learned to read eventually and things turned out okay (see, I wrote a book!).

The early-reading frenzy accompanying my kids' childhood began as a response to the perennial question: What do our children need for success in today's world? Parents want to raise healthy, happy, and empowered adults. But thanks to an unprecedented parental arms race — created by what William Deresiewicz calls the "artificial scarcity of educational resources" — we feel pressure for kids to get early-reading and enrichment activities and later to jump through countless hoops to impress college-admission committees.

What children really need to learn for their futures, aside from joyful rounds of "Baby Beluga," is how to be free and how to self-govern. For our youngest citizens, *that* requires parental modeling, exposure to government, and inspiring picture books.

My family knows three brothers who learned a powerful civics lesson early in life. The boys loved swimming and fishing in Oregon's pristine Metolius River, which their great-grandfather, a poet, had bequeathed to his descendants in a poem composed on its banks in 1921. When the boys learned that a proposed mega-development would use all the river's water and run it dry, their family traveled to the state capitol to support a proposed bill to protect the river.

The bill failed by one vote. The boys' mother remembers, "They left the chambers crying. They just couldn't believe lawmakers would let the river be dewatered. We told them, 'Never give up!' and our whole family went to the governor's office that afternoon."

A few weeks later, a new bill to save the river came to the floor, and this time, the eldest of the boys, third-grader Sage, told lawmakers, "It is very mean of the government" not to protect the river. "I may be only nine years old, but I should have the authority to have what people who are about forty years or older used to have when they were kids just like me."

That time, the bill passed.

"When children are in the room," their mother told me, "the entire conversation changes because children hold the moral authority."

Here are some wonderful resources parents can tap into — and suggest to teachers — to help us raise empowered citizens, starting even as early as in kindergarten. Trust me, they'll still learn to read.

If you have...

Fifteen minutes: Visit the Bill of Rights Institute (www.billofrights institute.org), which wants to "help the next generation understand the freedom and opportunity the Constitution offers." In the "Student Resources" section, download (for free) the kids' song "The Bill of (Your) Rights" and sing it together! Students in grades eight to twelve can enter the "We the Students" essay contest.

Enjoy author John Green's wacky thirteen-minute civics lesson: "The Constitution, the Articles, and Federalism: Crash Course US History #8" (www.youtube.com/watch?v=bO7FQsCcbD8&t=620s).

Thirty minutes: Anytime you're with more than one child, challenge kids — or adults, for that matter: Who can be the first to memorize the fifty-two-word Preamble to the Constitution? (See appendix E.) Offer a prize of your choice: homemade cookies? A paper crown? Twenty-five dollars in their college savings fund? Explain what words like *posterity* mean.

An afternoon: With your kids, visit the government spaces where decisions are made that affect their lives — and the health of their planet. Walk through your town municipal building and visit the mayor's office. Tour your state capitol. Some, like ours, have a kid-friendly scavenger hunt.

Similarly, attend key climate events, such as town hall meetings with your congressional representative, or local votes on environmental issues by your city council or county commissioner. If necessary, excuse kids from school for a real-life lesson on civics in action.

53. Empower Kids to Solve Local Problems

Why did we keep trying when all the adults said it was impossible?
Maybe because we're kids and don't know any better.

— ASHTON COFER, teen inventor

One day in 2009, two professors at the University of Oregon grumbled over lunch. Why is it, they asked, that countless young college students write great papers with innovative sustainability solutions every term, and no one but a professor ever sees them?

At the same time, city budgets keep shrinking — just when cities desperately need help with their sustainability problems in today's climate reality.

To build a bridge between youth hungry for real-world change and city officials seeking fresh solutions, professors Marc Schlossberg and Nico Larco launched a program that is now called Educational Partnerships for Innovations in Communities (EPIC).

Here's how it works: For one year, a university or college matches dozens of their classes with a city, which then treats the students as sustainability consultants. Hundreds of students from various departments — architecture, business, economics, public relations — are unleashed onto projects, such as designing zero-emissions police stations or launching campaigns to increase bike-walk-roll participation at local schools.

It's brilliant. Students report feeling happier, professors enjoy teaching students who are more engaged, and city officials appreciate the infusion of youthful energy into what can feel like a stale bureaucracy. For her bicycle transportation course, one student I met worked on making her city's bicycling infrastructure safer, more accessible, and family-friendly. For her "exam," she pitched her most innovative ideas to Redmond, Oregon, city officials just as she would to real clients.

"I put much more energy into my work because I was making a difference in the real world," she told me.

EPIC is so successful that it's already been replicated at twenty-six US colleges and universities and is quickly spreading abroad, including at twelve universities in southern Africa. The UN has seized on it as a promising model for global sustainability education, Schlossberg told me, because it's well-suited for low-income regions; it doesn't cost much to implement, since it works with existing higher education programs and city bureaucracies, and it requires only the addition of a coordinator, which is essentially funded by the fee the city pays into the program.

Why not replicate EPIC's model in K-12 classrooms and in our homes? Kids have big imaginations and like to solve big problems. The world could use fresh thinking on its big problems. Let's put them together, folks, right in our own communities!

Imagine the impact on students like my son. Forrest doesn't want to solve math problems just for the sake of it, but he might enjoy doing computations to design a bicycle bridge across the busy street he hates to cross twice every school day.

Or imagine if different classes in one middle school took on various ways to lower your city's carbon emissions by 10 percent: One student might try to reduce car idling, while another created a public service announcement about properly disposing of old appliances, and yet another studied the energy savings of government buildings using motion-sensor lights.

There are thousands of potential kid-engaging projects right under our noses.

We don't have to reinvent the wheel. Schlossberg and Larco give away their ideas every year at their replication conference — that's how new programs keep launching. So hey, school administrators! Go! Then help students blossom when their ideas are taken seriously.

The simplest application is, of course, in our families. We can let kids take ownership of sustainability problems in our homes, clubs, or churches, and really listen to, help develop, and try to implement their ideas. If the EPIC program is any indicator, it's energizing for adults sometimes trapped in outmoded thinking — and empowering for young future leaders learning how to heal the planet and protect their communities.

If you want to...

Challenge your kids: With younger kids, Schlossberg suggests playing the "electric meter game." Ask kids to watch the electric meter in continuous motion. Then challenge them to find out: How slowly can they make it go? How much energy is used when every appliance seems to be "off"? What's the impact on the meter as specific appliances (like computers and TVs) are turned on? How many can be turned off and kept off until needed?

Pitch to older kids: "If everyone reduces their miles driven by only 3 percent, it will significantly cut emissions. Can you figure out how to reduce our family's driving miles by 3 to 10 percent?" As necessary, help them log the number of miles the family drives weekly and monthly, and for what purpose. Then, have kids devise ways to reduce that number.

Challenge local colleges, churches, and organizations: As high schoolers look at prospective colleges, consider those that offer project-based learning, such as using the EPIC model (www.epicn.org). If preferred schools don't, email the provost to request it.

Ask churches and organizational leaders to task youth groups with solving real sustainability problems. How can they cut the monthly heating bill? Go zero-waste?

Or, as Schlossberg suggested to me, "Instead of your church working on one issue, the local synagogue working with another, and the local elementary school on another, try getting all three to focus on the same thing over a single year and watch how much more impact, engagement, and knowledge will emerge. Each group can do different things and don't even really need to coordinate much — just the fact of working in the same place over the same amount of time changes everything. Scale matters — that's the lesson here."

Be inspired: Watch teen inventor Ashton Cofer's Ted Talk, in which he describes his team's conversion of Styrofoam waste into activated carbon for purifying water (www.ted.com/talks/ashton_cofer_a_young_inventor_s_plan_to_recycle_styrofoam).

54. Facilitate Empowering Conversations

Ever feel like no matter what you say, or how you say it, your parents never really listen to you?...It's OK! You are normal.
— DEBRA FULGHUM BRUCE, "I Can't Talk to My Parents" on WebMD

When my son's best friend Abi turned twelve, he began bar mitzvah preparations. In the Jewish tradition, age twelve to thirteen is a time of great conversations, when young people, assisted by their rabbi, community, and family, examine their place in the scheme of things. Abi, adopted from Ethiopia at age ten, explored ideas about family and identity and also, because his late biological father had been abusive, about violence.

By contrast, my middle school rite of passage consisted of church confirmation classes that covered mostly Bible stories, after which, at an end-of-the-year church service, I unquestioningly accepted Jesus as my savior.

I didn't have a mentor as Abi did, nor — perhaps because my big family was less introspective — much conversation linking religion with the rest of my life. I honestly never understood my church and, being too self-conscious to tell anyone, I eventually left it.

After witnessing Abi's blossoming through his bar mitzvah experience, I wondered how, lacking church or synagogue, my teens might find meaningful places in an imperiled world — and an imperiled democracy. When Abi's mother told me about their in-depth dinner conversations, I decided to start there.

Shared family meals are increasingly rare in busy family life, but Art and I have always prioritized sharing conversation and nutritious food, so we were off to a good start. To facilitate connection, we continued our "no reading or phones at the table" policy.

To move conversations beyond everyone's stories of the day, we first tried store-bought conversation-starter cards, which were fun but shallow. When I worked harder to plumb deeper into my children's thoughts around classes, teachers, opinions, and philosophies, I noticed something important.

Art and I talked too much. I heard myself holding forth on political topics, trying to educate, I guess. My husband interrupted kids to interject his knowledge or correct insignificant minutiae.

No wonder they weren't sharing their thoughts at dinner. We weren't, I realized, always truly listening.

We tried harder, prompting them, "Tell me more" or "What do you think?" Then we listened — phones off, mouths zipped, eyes smiling.

That alone shifted my relationships, especially with my son. Being listened to with deep respect, interest, and patience facilitates conversations with him better than any artificial conversation-starter cards — and it creates more harmony in general. Now I ask just a few questions and create space for safe, respectful exploration — rather than mentally prepare my next brilliant comment.

Again and again, I'm reminded that I have so much to learn from young people. This is especially true when Abi and Forrest banter, exchanging wonderful ideas about where, actually, the universe ends, and where, if anywhere, God might inhabit it.

If you have...

Time but no money: To facilitate empowering conversations:

Fill your home with music, poetry, classic books, and movies from the library. Talk about these at meals. Read lyrics to your kids' favorite songs aloud; ask what they love about them, and find something to praise about the music or lyrics.

Over meals, invite stories from everyone's day. Share your own, but also ask, "What was one funny thing that happened?" or "What made an impression on you?"

Ask questions, and listen without judgment or — hardest for me — trying to fix anything. Being listened to by an adult can be a rare and empowering experience for kids.

Tap the wonderful Family Dinner Project (https://thefamilydinner project.org), which provides resources for "food, fun, and conversation about things that matter."

A little money: Stock bathrooms and dens with magazines such as *Skipping Stones* for children, and *Adbusters*, *Utne Reader*, or *Rolling Stone* for teens. Ask "What do *you* think?" questions over meals.

More money: Enroll kids in youth groups, wilderness retreats, or empowering creativity programs such as Power of Hope (https://powerofhope .org), whose mission is "to unleash the positive potential of youth through arts-centered intergenerational and multicultural learning programs."

55. Let Kids Play with Knives

Let them take risks, for godsake, let them get lost, sunburnt, stranded, drowned, eaten by bears, buried alive under avalanches — that is the right and privilege of any free American.
— EDWARD ABBEY

One summer day when Forrest was twelve, his friend Jake invited him to the woods for Spy Wars.

With airsoft guns.

In a heartbeat, I was on the phone, interrogating Jake's mom: *What kind of guns? Is that legal? Won't someone lose an eye?*

I desperately wanted to evict Forrest from the couch. He'd lounged too much that summer — granted, mostly reading — and I also was tired of our battles over screen time.

But...guns?

Jake's mom reassured me. "They're little pistols spring-loaded with plastic pellets. They have orange tips to signal they're toys, and the boys play in a remote wooded area. And wear goggles."

Hmmm, okay. Still...

"Mary," she said, "they run outside *for hours*."

Sold! Forrest exploded out the door and returned happily exhausted hours later. He bought his own airsoft gun and spy-warred daily until school resumed. No one lost an eye. Now, five years later, he throws knives, trying to stick them into dead trees whenever we camp.

In bygone days, kids played outside with sharp and dangerous things, roaming the natural world without adults hovering. These days, kids play inside with dull and dangerous things, roaming the digital world, often without adults noticing, much less hovering.

In the past, kids learned to use slingshots, fire, sledgehammers, axes, and knives to feed, clothe, and shelter themselves. Now, they watch people on screens do those things, rarely moving their own bodies or learning anything useful.

Global warming makes it more critical than ever for kids to become self-reliant adults, yet few parents allow unstructured outdoor playtime. The prevailing culture of fear compels us to pen our children inside, safe from pedophiles, tetanus, drowning, ticks, cliffs, melanoma, and accidents from trampolines, bikes, or, god forbid, tree climbing.

Research supports not only getting kids outside — some pediatricians even write "park prescriptions" — but letting that play be unstructured. Meeting the world directly, with no parental savior hovering, empowers children to learn independence and appropriate boundaries.

We took that to heart and resolved that, as our kids felt ready for a new challenge — such as a trampoline flip or solo bike ride — we'd give a thumbs-up. This hasn't always been easy — fear's grip is tight — but in twenty-one years, our kids have always been right about what they can handle (knock on wood). And buying a trampoline — over the objections of the pediatrician — was one of the healthiest decisions we made for our kids.

Now, our kids travel alone in foreign cities, challenge injustice, camp regularly — even in snow caves — and, thanks to their pyro-play (as we watched), they can build a toasty campfire to avert hypothermia. Their love of nature grows alongside their problem-solving skills, autonomy, and sense of aliveness, all key ingredients to healthy lives and democracies.

As your kids are ready, I encourage you to set them free in their natural habitat and teach them to thrive in it. The occasional bites, burns, and bruises will be small prices to pay for the empowerment.

If you want...

Advice and inspiration: Visit Lenore Skenazy's website Free-Range Kids (www.freerangekids.com), and read her book by the same name. Also check out: *Let Them Be Eaten by Bears: A Fearless Guide to Taking Our Kids into the Great Outdoors* by Peter Brown Hoffmeister and Richard Louv's *Last Child in the Woods*, which spawned the international No Child Left Inside movement.

To teach outdoor skills and self-reliance: Go camping (or rent a rustic cabin). Spend days hiking and evenings around a campfire, cooking outdoors. Fish! Let kids set up tents, build fires, gut fish, and whittle sticks for making s'mores. Or enroll kids in an outdoor program that teaches these things.

Climb big trees! Visit the Global Organization of Tree Climbers (www .gotreeclimbing.org). To climb high using gear, train with tree-climbing instructors (find one on their interactive map).

56. Fight for Climate Literacy in Schools

Climate education, especially in the age that we're living in, needs to be a whole lot more bold.

— TIM SWINEHART, teacher

Pop quiz: What percentage of science teachers teach our kids that global warming is real and largely caused by humans consuming fossil fuels?

If you were hoping for 100 percent, I'm sorry to report that according to a 2016 study, it's just over 50 percent. Worse, 30 percent of science teachers emphasize the idea that planetary heating is due to *natural* — not human — causes, and 27 percent give "equal time to perspectives that raise doubt that humans are causing climate change." A full 10 percent deny any human role. Half of US kids absorb climate science that's hit or miss at best — and often wildly inaccurate.

The good news is that most teachers want up-to-date information, support, and effective teaching materials. The bad news is that the Heartland Institute — a well-funded conservative think-tank — produces and sends glossy brochures to teachers that explain how to teach our kids that global warming may not really be happening or, if it is, that "many areas of the world would benefit."

Thank heavens many educators, parents, and school boards are waking up to this manipulation. In 2016, Portland's school board unanimously passed a first-in-the-United-States resolution to implement a climate justice curriculum and reject textbooks downplaying humanity's planet-cooking role. Other cities are seeking to replicate Portland's model.

It's deeply unethical to misinform young people about the greatest crisis they face as a generation. Moreover, climate education focused on do-able solutions proposed by the world's top climate scientists — keeping fossil fuels in the ground, cutting emissions, switching to clean energy, planting trees, and doing it *now* — energizes and empowers our children.

Tim Swinehart's Climate Justice class at Portland's Lincoln High, for example, teaches the science as well as how to write speeches, testify at public hearings, and engage public officials. One of his classes helped pass the first municipal climate-recovery resolution in the U.S. that blocks new fossil-fuel infrastructure. Seventeen-year-old Tyler Honn told *Yes!* *Magazine* that the class has given him "much more confidence, a sense of agency, and a sense of purpose that I absolutely didn't have before."

I asked my son's former fifth-grade teachers, Susan Dwoskin and Carrie Ann Naumoff, how they handle climate literacy in elementary school. They said they study the Bill of Rights and environmental sciences, and every day, year-round, they weave in the climate justice concept that "everything is connected to everything else."

"Many people say, 'We can't put that fear into children. It's too much,'" Susan told me. "*We're* saying there are appropriate ways to approach climate justice. Children really want to know what the real issues are." The key, Susan said, is not to teach students what to think, but how to think critically.

"We read information and learn how to get the best source materials possible," Carrie Ann said. "We ask, 'What's the author's perspective here? What do you get out if it? How does it compare to other authors' thoughts on this issue?' The kids make the connections."

What the teachers see, year after year, is growth in overall student confidence. Carrie Ann said, "In the first two weeks of school, we ask, 'What do you care about?' We open up the floor to the kids, and pretty soon there are kids who want to save sea turtles or whales or an uncle's job as a logger. *They* choose, not us."

Then, civic engagement kicks in. Past fifth-graders have presented their climate research to visiting high school science teachers, to the whole school, and to public officials. In the chapter "Sue the Grown-Ups!" (page 244), I describe fifth-graders making handmade posters in support of Kelsey Juliana and appearing on national TV. These were Susan and Carrie Ann's students. That experience — engaging with real-world decision-making and speaking in public — was a school-year highlight for them.

Now more than ever, engaging children about climate justice is especially critical, and below I provide Carrie Ann and Susan's advice for taking charge of this yourself.

If you want to...

Push for climate literacy in your school: Carrie Ann and Susan offer this four-point plan for climate-conscious parents:

1. Talk to other parents. Have allies first. An individual parent or teacher might not effect change, but parents as a group have a strong voice if they go together to parent councils, PTAs, site councils, principals, and school boards. It has to be a ground-up movement.

2. Bring kids. Kids presenting at school boards is profound — and it impacts kids, who suddenly want to have more of a voice.

3. Find out who decides curricula for your school and go to those meetings. The site council at our school meets several times a year and functions like a board; everyone helps decide what's taught.

4. Fight to keep local decision-making power over classroom curriculum. Our district administration is currently trying to take away our site council. Many don't want teachers and parents to have this kind of power, but we're fighting for it. Be ready to keep fighting.

Supervise what your child is learning: Look at your kids' science textbooks. Ask to see science teachers' curricula. Talk to your child's teacher and school if either of these important educational resources cast any doubt about the reality and causes of global warming.

At parent-teacher conferences, ask how climate change is included in the daily curriculum and request that it be expanded. Write this in a letter to the school, and request that any "info packets" from the Heartland Institute be rejected. As Susan says, "Support teachers teaching good climate science. And don't ask, 'Did you meet the standards?' We actually stretch the standards to fit our curriculum."

Finally, attend key school board meetings when important district-wide decisions are being made about climate education. Watch parent newsletters for announcements.

Discuss issues at home: Susan believes that climate literacy starts at the dinner table. "Put your phone down and talk to your kids," she says. "Teachers send home what students are working on. Look at it, talk about it, critique it together."

No matter what is taught in school, you can raise climate literacy at home just by keeping in mind Carrie Ann and Susan's basic framework: *Everything is connected to everything else.* Remind your children of that every day, and use the information and advice in this book.

57. Unplug the Kids

There is compelling evidence that the devices we've placed in young people's hands are having profound effects on their lives — and making them seriously unhappy.
— JEAN M. TWENGE, psychology professor

In old Ireland, Mam and Da worried about Stranger Danger just as we do. Only they feared the Little People, fairies who steal human babies and leave behind glassy-eyed changelings with a ravenous appetite for cake.

Which sounds like the teens I found on my couch one sunny afternoon — pale, covered in cookie crumbs, and staring dully at their phones. I tried the usual fairy-banishing remedies, but neither baptizing them nor nailing an iron horseshoe above their heads revived them.

I did what any parent would do when confronted with changelings and signed them up for camp. No service in the wilderness! I also bought a router that limits internet time, which helps us prioritize books, movement, and family.

The battle continues, though, and our high school, ironically, has been my greatest foe, since online homework provides the perfect cover for nightly screen-fests. Unless we hover, we can't tell whether they're doing research and schoolwork, posting cat videos, or falling prey to click-bait. Neither can teachers, apparently. Kids now watch movies on their phones *during class*. Many teachers even ask students to access the internet on their phones for classwork, or award points for posting on the class Facebook page, normalizing an unhealthy trend that may explain how, in 2015, kids spent an average of *nine hours a day* on-screen.

My husband and I refuse to abandon our kids to the predations of the entertainment industry. We allowed no screen time — even movies — before age seven, then we introduced it in limited doses as they grew. We delayed gaming until fourth grade, and then fiercely held total daily screen time to twenty minutes (plus occasional movies) until age sixteen. No killing games allowed. That took determination and was easier to do a decade ago, but I would do it again to protect their health.

We also provided lots to do — hands-on games, punching bags, dartboards, a tree house, and more — and an open door for friends and neighbors. When off screens, our kids seemed happier. They were definitely nicer.

How can parents unplug kids?

First: ignore studies promoting screen time. Seriously. Cover your ears and "la-la-la" if you must when friends quote them, since tech entertainment companies fund many of these studies to confuse parents and keep everyone buying their products. Just like tobacco companies did. Just like Exxon does. Misinformation works. Kids need movement, reading time, and free play.

Second: Befriend screen-free or screen-limited families. This makes child-rearing easier.

Third: Trust that screen-limited kids aren't disadvantaged later, as many parents fear. My kids had little screen time before high school, and they're tech proficient, getting good educations, and gainfully employed. They're fine. Yours will be, too.

Go ahead, delay those damned screens, then limit them. I promise, your kids will thank you someday, too.

If you have . . .

Kids in school: With another parent, ask your child's principal to limit online homework and Facebook groups for school or team information. Bring back handouts and parent emails! End phone use in classrooms!

Kids under seven: Avoid screens, including movie theaters. Ask family, babysitters, and other parents to support screen-free activities.

For car trips: Try frequent breaks, kid music, riddles, finger puppets, window suction-cup toys, the I Spy game, the Alphabet game, twenty questions, and audiobooks (like *Seal Maiden: A Celtic Musical* by Karan Casey).

Kids under seventeen: Delay and limit gaming. Avoid kid smartphones and data plans until college (when kids pay).

Read the 2016 study "Effect of Early Adult Patterns of Physical Activity and Television Viewing on Midlife Cognitive Function" (https://jamanetwork.com/journals/jamapsychiatry/fullarticle/2471270).

Kids at home: Use routers like Disney's Circle to limit screen time; program it to pull the plug after a certain number of minutes daily, to switch off before bedtime, and to filter content.

Make it easier to be off screens than on. Designate screen-free zones and times. Stock your home with books, magazines, games, sports equipment, art supplies, alluring cookbooks, and friends.

Read *Wired Child* by Richard Freed or *Reset Your Child's Brain* by Victoria L. Dunckley.

58. Amplify Kids' Voices

In the climate issue, it's hard to make a kid your enemy. Kids hold the moral high ground.

— My friend Eliot Schipper

On Thanksgiving weekend in 2015, the three leaders of our local high school's Earth Guardians/350 club faced a crowd of people gathered to demand a binding global climate treaty from world leaders meeting in Paris. Fifteen-year-old Wesley Georgiev took the mic.

"We know we have a right to clean air!" he shouted to the hundreds of listeners gathered at the county courthouse. "We know we have a right to clean water. And we know we have a right to a positive future!" As people cheered, he handed the mic to the next speakers, Avery McRae and Hazel Van Ummersen, two girls suing the U.S. government for destroying the climate and violating their constitutional rights.

"Honestly, I wish we didn't have to spend our time doing this," said Avery. "I'm a ten-year-old girl who likes riding my bike, playing with friends, listening to music, and just having fun! But adults are *simply not* doing enough about this issue!"

Then the kids taught the crowd a chant in which the kids called, "Kids who want a future/are amping up the sound! Let us hear you shout it..." — and the crowd joined in — "Keep it in the ground!" From the head of the march, the kids kept the call-and-response going all the way through downtown, which was packed with holiday shoppers, and into a riverside park. I then led the crowd through an interactive art project — a 450-person human oil drip that transformed itself into a bright sun as we filmed it from above — after which several kids were interviewed by reporters for the evening news. That night, Wesley said to his mom, "This was the best day of my life."

What I've learned during several years of trying to mobilize the public for climate action is this: People want to hear from the people most impacted in their community — which, in this case, is kids. For their part, kids like to give adults a piece of their minds. Many kids, including my own, are bewildered as they witness adults destroy the climate. They ask fearfully, "Don't adults care what happens to us?"

Perhaps one reason it's been so hard to build a climate movement is that we've gotten this part wrong. The public we're trying to mobilize doesn't want to hear from people like me, or angry young white guys, or even scientists — not really. Most people don't want to, anyway. They want to hear from the people suffering the most, including our children, who can't vote yet but who will be left in the wreckage after we're gone.

In this kids-and-their-audience recipe, parents are the secret sauce. Most of the stuff that happened at the 2015 event wasn't organized by the kids. They wrote the speeches, built the buzz in their media worlds, and led the pack that day. They brought energy, sparkle, and adorable posters. But their parents and 350 Eugene volunteers wrote the press releases, built the giant puppet, rented the park, hired a videographer for aerial footage, brought first-aid kits and medics, charged batteries for the megaphone, and so much more. It was similar to any film or theater production I've worked on: Dozens toil behind the scenes, but when the show begins, all audiences care about are the actors.

This doesn't mean parents have to throw a big climate march to give kids a platform for expression. Many parents find it satisfying, though, to help kids do *something* — and if there's one thing my parenting generation has down, it's kid advocacy and logistical support.

First, brainstorm with kids. Listen to what they love to do, and think about upcoming events or holiday tie-ins. How about a "kid climate cheerleaders" squad outside the football stadium, happily performing chants and dance moves to cheer on the climate after the big game? A kids' climate art show on the walls of your favorite bakery? An entry in the annual city parade? A teen poetry slam-on-the-climate-brakes night? Organize one event — even just one humorous short skit for one talent show — featuring kids' voices. Give it a try and see what happens.

Wait, some ask. *Isn't this "manipulating" children for our own political agenda?*

I think I'll let a young person answer this one. After her court appearance for her climate lawsuit against the federal government (see "Sue the Grown-Ups," page 244), nineteen-year-old Kelsey Juliana agreed to speak with a reporter. Midway through the interview, the reporter asked if perhaps Kelsey was being used for adults' political agendas. Kelsey, tired of this line of questioning, fired back: "You're saying that because you're afraid to give young people credibility. You're nervous about our power because we're informed, upset, and taking things into our own hands. That's intimidating. So instead of listening to us, you try to shut us down." When the article appeared, the reporter shared that part of the story — and her regard for Kelsey's passion and power.

If you have...

Three minutes and twenty-eight seconds: With your kids, watch the video of kids speaking and leading the 2015 march described in this chapter (at www.marydemocker.com). My website also includes other videos to inspire young activists.

Fifteen minutes: Brainstorm with kids about what matters to them and one simple way they'd enjoy expressing themselves. In his book, *We Rise*, Xiuhtezcatl Martinez describes planting sunflowers outside a coal plant with his local Earth Guardians group. After the coal company yanked out the flowers and left them, the kids returned to give the sunflowers a funeral, which got as much press as the original protest.

To get attention from the local media and others, send out a press release. For an example, see appendix C (page 260), or look for easy online tutorials (such as on wikiHow).

An upcoming rally or climate event: Encourage your kids to attend. Help them raise their voices and make art or posters. My daughter, Zannie, is now twenty-one, and here is what she recommends: "Give kids snacks. If possible, give teens credit for some of the volunteer hours they need to graduate. And make it fun! It can be about something heavy, but give us something fun to do to help."

59. Permeate Kids' Worlds with Solutions

Focus on the solutions instead of the problem; when you do this, you have the ability to stay positive in your mindset, which is much healthier than being drawn down the rabbit hole of negativity.

— EARTH GUARDIANS

One fine fall afternoon in 2008, two dozen Girl Scouts surrounded me on my front lawn. One of their troop leaders, my friend Jennifer, had asked if she could bring the group of ten-year-olds to experience my "Personal Power Station."

This was my art installation in the lead-up to that year's election, which was a bit of a close shave for my candidate. I'd hoped to inspire people to leap into the lackluster affair by reminding them that the outcome was in our hands and actually mattered. My signage invited visitors to "Get Charged Up Here! Vote — And Get Your Friends to Vote!" To that end, a bright-red mailbox staked into my lawn stood ready, stocked with voter registration forms that, in my zeal, I'd even prestamped.

Dozens of jewel-tone plastic pouches, attached to various branches and suspended bars, hung at Girl Scout eye level. Inside each pouch were three-by-three-inch paper squares visitors were invited to take home; these contained inspiring quotes like Elie Wiesel's famous statement, "The opposite of love is not hate, it's indifference."

The girls toured the installation, each pocketing several quotes, until Jennifer gathered them for a Q&A with me about how ordinary people can change society. She got them brainstorming things they could do themselves, like make sure their parents were planning to vote or hanging "Vote!" signs in their windows. Finally, Zannie, then fifteen, helped them press apples from our tree into cider. This was a nice hands-on activity that gave them a treat and us a chance to talk about local volunteers who collect fruit that would otherwise rot under countless trees around town and donate them to food banks.

Looking back, I appreciate what a rare and cool thing it was for Jennifer to initiate and organize that field trip. So many of us parents dedicate energy and money to helping our kids gain personal mastery of violin, basketball, and homework after school. Yet, how many of those minutes in a week, or even a year, are spent connecting the dots between our kids' activities and climate recovery — the most important challenge they'll face as a generation?

For that matter, how much time is spent even *acknowledging* that the planet is warming, much less empowering kids with the necessary skills to turn things around fast?

Starting today, let's boldly highlight solutions everywhere in our children's world, making positive climate solutions the lens through which we view every activity, both at school and at home. Everywhere kids go, we can empower them with big-picture thinking because, for better or worse, every crew team, culinary club, or choir rehearsal is affected by — and affects — global warming.

Here are some starter ideas for connecting the dots and immersing kids in solutions. Not every coach, choir director, dance teacher, or 4-H leader will be on board.

But remember, to create a successful movement for change, only 3.5 percent of them need to be.

If your child is in a . . .

Scout troop: Propose a scout trip to a wind or solar farm to learn about the clean-energy boom sweeping the nation. Help organize transportation.

Dance class or troupe: Suggest that dance teachers choreograph one dance about wind farming or tree planting — and protection — for the spring recital. For inspiration, share this music video of the Bay Area Youth Harp Ensemble advocating for the protection of California's redwoods (www.youtube.com/watch?v=-G_eRrZKvkM).

School club: Challenge the robotics club to use 100-percent renewable energy to power their robots. Donate parts, if necessary.

Suggest the bicycling club or church youth group spend one afternoon riding to city hall to advocate for well-lit, tree-lined bike lanes and secure, covered bicycle parking. Help organize logistics.

Ask drama teachers for plays about social struggle and environmental (in)justice, such as *The Lorax* or *Earthlings*. Or help — or find someone to help — kids write and perform their own play.

GROW
COMMUNITY
CONNECTIONS

60. Set the Bar Low for Allies

If you really, truly, in your heart want change, you have to make everybody your ally.
— RICHARD GREEN, Crown Heights Youth Collective

I knew only a few of the thirty or so people at the stop-the-pipeline action meeting, so I was glad when the leaders — hardworking volunteers — announced an icebreaker. We stood in a circle, instructed to step forward every time a statement applied to us. The facilitators prompted:

"You've joined antinuclear protests.... You don't own a car.... You're vegan." In and out of the circle I went a dozen times, mostly staying out and feeling increasingly self-conscious — and irked by the competitive undertone of the exercise.

When one leader said, "You've moved your money from commercial banks to local credit unions," one participant, another hardworking volunteer, joined the inner circle, then wagged a finger at those in the outer circle, admonishing, "You re-e-e-ally should!"

Great. Now some of us felt ashamed.

That wasn't anyone's goal, of course, but moments like these help explain why environmentalists are sometimes accused of being elitist jerks. Our passion can easily morph into self-righteousness and even dismissal of entire demographics — especially of those busy working two jobs, caring for others, lacking in resources, or sometimes all three.

At recent political events I've heard frustrated activists say things like, "Families aren't politically active because they're busy living the American Dream," or "Rednecks won't give up their monster trucks." I've had my share of blaming thoughts — they're difficult to stop sometimes, given our crisis. But neither of those statements is actually true, and this line of thinking is a trap. Countless families and "rednecks" care deeply about the natural world. We're all captive, regardless of whom we voted for or how many monster trucks we own, within a dirty-energy infrastructure. This bears repeating:

We're all captive within a dirty-energy infrastructure.

It's not our individual shopping lists we must change. It's our Stupid infrastructure.

That's a huge job. We're fighting corporate power, political corruption, and human inertia. We may not win. So, for heaven's sake, let's not challenge one another at the door and villainize one another for our polyester, pickups, burgers, or banks.

Let's stop "should"-ing on ourselves and others, and that includes the facilitators I describe above, who were volunteers kind enough to try leading an icebreaker. Let's just do our best to change things from Stupid to Smart as we work — and play — side by side. We can begin with the assumption that everyone wants a livable planet, and stop requiring, however subtly, that individuals be lifestyle superheroes.

Bernie Sanders does this well. No matter where he goes — and he regularly visits conservative communities — he connects with people around basic human decency. Few people actually disagree with his belief that we all deserve a living wage, health care, and affordable education, so he works from that common ground. We can do the same around climate issues. Almost everyone wants clean air, water, and soil. That's a great start.

Keep the bar low. Take responsibility for good communication. I honestly don't care whether you drive an electric car or a Hummer. Wanna help save the planet? Step forward, into the circle. We're so happy you're here.

If you want to join forces with more...

Families, immigrants, students, lower-income populations, or any political party or groups with which you don't identify: Go to their spaces and events. Listen to their concerns, and how climate destruction particularly impacts those concerns. Ask how to be supportive; ask what kinds of events and times work for them. For students, hold some events on campus. For parents, provide childcare and snacks. For non-English speakers, make sure your announcements are in any other languages used widely in your area.

Mentor others on speaking out effectively. That may mean providing logistical support and handing over the mic to them.

Advertise that events with admission fees or that request donations are on a sliding scale (or free for students). Evaluate the culture you or your group has created (and each group has one). Is it cliquey or truly inclusive of everyone?

61. Have a Beer with Cousin Max

Refocusing the climate change conversation on life as the reason we're fighting will serve to bridge the political divides that have impeded our progress.
— JAMIE BECK ALEXANDER

The teens at the table suddenly got very quiet. At our usually boisterous family reunion dinner, a beloved relative began to hold forth on geological warming cycles, ice ages, and the idea that humans aren't causing global warming, or if they are, it doesn't really matter because "natural warming cycles dwarf any impact humans might have."

I felt sick. *When did* he *start buying this crap?* Though our relative has no training as a scientist, much less a climatologist, he spoke with such authority that I actually began to wonder if maybe he was right. Was it possible *I* was the one who wasn't up-to-date on the best science?

The teens caught each other's eyes, and my kids' eyes flicked to mine. I felt the silent expectation that I should take him on, since I write about this stuff — and we were in too deep to change the subject. I girded myself to challenge him diplomatically, but I couldn't find calm, authoritative words.

As he held forth, I nodded politely while calling a little side huddle with myself:

Remember self-help guru Dr. Wayne Dyer's suggestion for tense moments like these at family gatherings? Just ask open-ended questions. Try to find out why *he thinks the way he does.*

Feeling calmer, I asked, with genuine curiosity, "Do you believe, then, that we shouldn't try to lower our carbon emissions?"

"Oh no, we should do something!" he said. Everyone relaxed, and the teens began to participate again, gently stating their convictions — and fears — about the climate.

It was a watershed moment. For years, I'd either fled these conversations with "Cousin Max" (my euphemism for any beloved naysayer, including friends) or gone to battle, only to later kick myself for wasting my precious time or not having enough of just the right scientific factoids at my fingertips.

I realized that if I didn't try to convince our relative, I could relax. My real role was to model for my teens how to disagree with someone I love, see rarely, and hope will return to my home next year.

Letting go of being right allowed me to share my own fundamental truth. I said to our relative earnestly, "I would die for my kids. I want to keep them safe, so I'm volunteering nearly full-time for climate recovery. I'll keep doing everything I can until a group of truly expert climatologists convinces me that my kids are safe."

"Yeah, that makes sense," he said, nodding thoughtfully. As we all ate lasagna and adults drank beer, he softened his position, and in the end, he even agreed with some of what we said.

I've found that, in private conversations, people respect my parental protectiveness angle, regardless of their political bent. Everyone in our close circle loves my kids and wants them safe. Seeking that common ground — especially with someone in our family — often results in an authentic exchange. It pivots the conversation from partisanship toward something deeper.

If that fails, I try to exit fast with my heart still light and determined. If people deny what's happening — whether out of fear, fundamentalism, or foolishness — then it's best not to waste my time or deaden my spirits with fruitless debate. Better to keep the relationship intact and choose a different battle — or none at all.

If you have...

Five minutes: Read "Learn the Fossil Fuel Nitty-Gritties" (page 66), which succinctly explains the climate problem — and solutions.

Twelve minutes: For those who like science-oriented approaches to those impromptu global-warming chats, use the bountiful, excellent resources from the Yale Program on Climate Change Communication (http://climatecommunication.yale.edu), including up-to-the-minute climate research and a daily radio show. Or visit Skeptical Science (https://skepticalscience.com), which offers several "debunking" tools — including a handy iPhone app. See also "Inoculate Everyone" (page 230).

Forty minutes: Write a script with kids for talking with "Cousin Max" or the *Fox News* devotee next door. Rehearse it — in costumes.

62. Chainsaw the Fence

Why buy a trampoline when you can just go jump on the neighbors' for free?
— ANDREW THOENNES, our neighbor, age eleven

As I write this, it's ninety-four degrees, and a kid party rages outside my studio. The neighbor boys and my son, Forrest, are playing in the pop-up pool that our neighborhood bought eight years ago. It lives in our yard because we have the most space, but our neighbors keep the paddleboard all winter in their (bigger) garage. Now, the boys are whooping with delight as they splash, jump, and wrestle on the paddleboard.

This communal scene began nine years ago when Jen and Steve and their three kids moved next door. Our two families' kids, wanting to play together, hoisted themselves over the wooden fence daily, until Jen finally asked, "Should we just cut a hole?"

Within minutes, my husband, Art, produced a Sawzall, and Jen did the honors. That afternoon, Steve got a load of wood chips and laid a path from their back door to our trampoline.

We had no idea how much this fence opening would change our lives. The countless easy playdates alone have been a great blessing — no calendars or car seats needed — but there is something else as well. Our friends Mimi and Nir (who'd encouraged us to move to the block years before) also live next door. Three close family households in a row means many eyes watching out for kids and for our homes when we travel. It means exercise buddies, easy carpooling, and home-baked desserts passed back and forth.

And we share lots of stuff, which means not only saved money but more neighborly interactions. We have a collective lawn mower, sauna (fired in part with pieces of the dismantled fence), tree house, canoe, basketball hoop, portable fireplace, and trailer. We share chickens, power tools, and countless other things that free up closet space and save the planet from more stuff (and the carbon pollution of making it).

A fun side benefit of sharing is that kids can be the couriers when someone needs to return, say, hair clippers. This often results in spontaneous gatherings and games, which increases everyone's health and happiness, sometimes giving parents more precious pockets of time to spend on planet-saving.

If you have...

A city apartment: Introduce yourself to your neighbors. Leave a welcome note with your contact info — and maybe cookies. Put May Day flowers on doors. Try holiday caroling.

Start a dinner club with three other families; each family hosts dinner once a month.

Make your apartment a kid magnet: Dedicate space in one room for screen-free play. Stock it with imaginative, interactive, age-appropriate items: puzzles, board games, masks, costumes, kazoos, hacky sacks, small Nerf guns, craft supplies.

A front yard, porch, or backyard: For a no-fuss gathering, host a warm-weather Bring Your Own Supper picnic. Name the time and porch, and people bring their own plates of food. We love this easy way to dine with neighbors.

In a covered outdoor bin or under a patio, stock items to inspire outdoor play: include dartboards, chalk, balls of all types, cones for goals, skateboards, scooters, or our all-time favorite, Hunger Games "boppers." Kids make foam shields and "weapons" from reclaimed PVC pipes wrapped in used pool noodles and duct tape (and then battle for hours).

A new neighborhood: Introduce yourself to your neighbors. Also, check out Mike Lanza's book *Playborhood: Turn Your Neighborhood Into a Place for Play* and his blog (http://playborhood.com). And do some of the following:

Buy from kid lemonade stands.

Offer clothes and shoes your kids outgrow to parents of younger kids.

Shovel the neighbor's walk, or ask your kids to (offer them cocoa upon return).

Check on elderly neighbors.

Encourage your kids to help adults unload groceries.

Hire neighbor kids to rake leaves and walk dogs.

Build a Little Free Library in front of your home (visit https://little freelibrary.org for tips and inspiration).

Sawzall a hole or doorway in your fence.

No sense of neighborhood: Consider moving close to another family you love. It was one of the best parenting decisions we've made.

63. Host a Disaster Party

Remember: In a disaster, your most immediate source of help are the neighbors living around you.
— MAP YOUR NEIGHBORHOOD PROGRAM

One night, in the middle of an ice storm, a tree fell onto our ninety-year-old neighbor's power line. My husband immediately knocked on her door. Sure enough, a jolt of electricity had sent her flying, and she lay injured in the darkness next to a dead phone. She was deeply relieved to see Art.

That spring, I received an email from the environmental group Stand .earth informing me that my family sleeps inside a "blast zone." Should one of the oil trains now rolling through our town at night explode, as one did in 2013 in Lac-Mégantic, Quebec, killing forty-seven people, the few blocks between us and the tracks probably wouldn't protect us from toxic smoke and fire. Firefighters would be overwhelmed, and we'd be on our own.

A year later, the *New Yorker* published an in-depth article about the major earthquake/tsunami that is long overdue to hit the Pacific Northwest. FEMA spokesperson Kenneth Murphy said, "Our operating assumption is that everything west of Interstate 5 will be toast."

Hey, *we're* west of I-5! The article warned that, if this double whammy occurred, nearly thirteen thousand people might die and twenty-seven thousand be injured. Millions would be desperate for food and water.

I decided to gather the households closest to us to discuss disaster preparations.

One spring evening, ten of us shyly gathered. After breaking the ice with some neighborhood news — a raccoon living near our shared chicken coop; the kids' lemonade stand planned for that weekend — we got to work, using a tool from the City of Los Angeles called "5 Steps to Neighborhood Preparedness." This helped us to

1. identify our neighborhood,
2. choose a leader,
3. identify neighborhood threats,

4.   share contact info, and

5.   designate a neighborhood gathering place.

Item 3 inspired the most conversation. We discussed the likelihood of all three upstream dams breaking in an earthquake and flooding us. Where would we run? We talked about a possible oil-train explosion, which gave me a chance to educate neighbors not only about the oil trains secretly running through town, but 350 Eugene's work to stop them. We talked about climate impacts — the fir tree die-offs, the lengthening wild-fire season, and climate refugees — in a nonthreatening way.

We also learned who has CPR training and laughed a lot. One neighbor displayed his stash of silver dollars, in case civilization really collapses, and walked us through his prepper supplies, which was fascinating. Did you know you can order, for seventy dollars, a bucket containing 390 meals? Or that a "bathtub bladder" lets you quickly store a hundred gallons of water?

Most of all, we got to know one another better. People appreciate neighborliness, and we've always found that people are willing to overlook political differences in the interest of shared goals — like the safety of our animals, property, water supplies, and family members.

We made disaster prep shopping lists together — mine includes a water barrel, that meal bucket, and, yes, a tub bladder — and now we have a text group we test regularly. And because we're facing the idea of a crisis together, whether it's an earthquake or climate-related catastrophe, we have a neighborhood that feels a little more like family.

If you have...

Five minutes: Read your city's emergency preparedness page. Make a list of personal emergency supplies to gather. The Red Cross has a good list and a family disaster plan worksheet (www.redcross.org/get-help/prepare-for-emergencies/be-red-cross-ready/get-a-kit).

Subscribe to newsletters from your neighborhood association, if you have one.

An evening: Host a neighborhood emergency preparedness party. The Los Angeles five-steps program takes about ninety minutes (http://5steps.la).

Have someone volunteer to tackle the Map Your Neighborhood program (https://mil.wa.gov/emergency-management-division/prepared ness/map-your-neighborhood). This creates a plan with ten to fifteen households and a map of resources, skills, and vulnerabilities.

Thirty hours: Take the federal Community Emergency Response Team (CERT) training (www.ready.gov/community-emergency-response-team). Volunteers are trained in disaster preparedness, fire suppression, disaster medical operations, and search-and-rescue operations.

64. Kick Out the Climate Blues

If I can't dance, I don't want to be part of your revolution.
— EMMA GOLDMAN

I didn't know when I'd last felt so low. Billionaire climate deniers were ravaging my kids' future, a close friend had just died, and the downpour outside was forecast to continue another ten days. I wanted to beam myself into a sunnier, happier world — without the plane ride.

Then, Forrest plopped down to do math and cranked up a "get pumped" hip-hop song, "Rapper's Delight." The next thing I knew, I was dancing around the house, feeling pretty fine.

My change of mood shouldn't have surprised me, since I'd just witnessed the transformative power of dancing at my town's January 2017 Women's March. Thousands of us were feeling angry, grief-stricken, and frightened by Donald Trump's presidential inauguration, and so we marched together through downtown. A drum troupe waited at the end of the march, and I watched as hundreds of people — including my own teen — joyfully danced together in the streets to live music, despite the chill in the air and in much of the nation's mood. Forrest later shared with me that, in that moment, he felt more hope and determination than he had in months.

A 2015 Oxford University study explains why: When we dance with others, our bodies release bonding hormones and endorphins that elevate our health, moods, and sense of social interconnectedness. Isn't science wonderful (and, on a side note, important to listen to)?

So, beautiful people, until we remove the crazies unleashing more nightmares daily on our planet, we'll need to remind one another of this:

When you're blue over the state of the planet — or for any reason, really — turn up the music, roll up the rug, and grab the kids for a living room dance party. For maximum benefit, sing the words at the top of your lungs — even if you have to make them up.

If you have...

Young children: Usually all it takes to get young kids moving is a little fun music. Try cranking up some Putumayo-produced world music for kids (www.putumayo.com), such as *Kids World Party*. This provides a sweet entry into another culture and brings children's awareness to other people in faraway places who are worthy of our kindness and respect.

And try one of these songs:

"A Beautiful, Beautiful World" by the StoryBots
"Supercalifragilisticexpialidocious" from *Mary Poppins*
"Baby Beluga" by Raffi
"Octopus's Garden" and "Here Comes the Sun" by the Beatles
"Dad's Dunny" by the Formidable Vegetable Sound System (http://formidablevegetable.com.au)

Older kids: Ask your teen to catch you up on their favorite music. Challenge them: "You choose five songs for me to hear, and I choose one for you." Make it a fun intro to intergenerational music, old classics, and maybe even their grandparents' music. My music-loving teen and I bond over his favorite songs by Macklemore, Red Hot Chili Peppers, and Metallica, as well as my old favorites, which include Don McLean's "American Pie" and Aretha Franklin's "Respect."

Your sweet self: Tune in a music streaming service and try their feel-good stations or create your own. On days when I'm reeling from bad climate news, I pick myself up with Michael Franti's life-affirming song "I'm Alive (Life Sounds Like)." For a list of other songs that get me rolling up our carpet, peek at my website (www.marydemocker.com).

65. Draft Grandma and Grandpa

We elders are at the peak of our ability to help. We have a wealth of experience. Many of us have sufficient income. And we have that huge commodity: time. Most of all we have a ferocious love for our grandchildren.

— KATHLEEN DEAN MOORE

When my niece Heather started elementary school, she was shocked to see throwaway Styrofoam trays in her lunchroom in Takoma Park, Maryland. With help from her mom, a zero-waste expert, she helped lead her school's Young Activist Club in a quest for reusable trays. When they had to meet with local lawmakers to make it happen, Heather's whole family showed up at the town meeting in support.

"It was a very moving evening for me," Heather's grandmother, Marlene, said. Seeing the students' passion inspired Marlene's first-ever letter to the editor, published that week in the *Washington Post*. She went on to join every Young Activist Club event, even gathering "ban Styrofoam trays" signatures at the county fair alongside her granddaughter.

I hear stories like this more and more lately. Grandpas cut dozens of large cardboard sunflowers for high schoolers to carry in marches. Grandmas block oil trains while holding poster-size photos of grandchildren. Grandparents gather friends in senior living centers to collar elected officials or, watching grandkids suffer from asthma, fight to close the coal plant near their home or school.

Here's the thing. Grandparents adore their grandkids. They get to spoil them and brag about them, but (usually, anyway) they don't get stuck with the night shift or cell phone bill. However, as grandchildren become teens, and spend more of their time with friends or love interests, grandparents can't always find ways to participate in their grandkids' worlds. One way to bridge the generation gap is to invite elders on our planet-saving mission.

In fact, why not throw the invitation wide? Invite your siblings, aunts, uncles, and neighbors whose own grandchildren live far away. People usually appreciate invitations to participate in group efforts, whether they're climate-conscious or not. When my kids drew two hundred postcards asking President Obama to reject the Keystone pipeline, their aunt

Sharon was delighted to receive a package of three. She texted, "They're beautiful! Thanks for inviting me to help!"

We can also include those who are housebound, even temporarily, by disability, illness, or injury. When I was nineteen, I spent a lonely summer in a body cast and mostly in bed. I would have benefited tremendously from being part of an important campaign with my pen or phone. Feeling needed, after all, is a fundamental human desire.

We won't save the world alone. Let's ask for help, confident that some invitees, especially grandparents and extended family members, will be delighted to join the fun — and appreciate being part of something so important.

If you have . . .

Grandparents or extended family: Ask elders to join your family in your efforts to "Hound Representatives" (see page 178). Suggest they load their phones with representatives' numbers or the 5 Calls app (https://5calls.org). Offer tech help — from you or your kids — if grandparents need or want it.

Encourage retirees to attend daytime town halls, public hearings, or key votes that are difficult for parents to attend, but where public input really matters.

Invite them to choose one action from this book they might like to try with you and/or your kids.

Suggest grandparents sign up for a climate newsletter your family receives, then discuss it over Sunday brunch.

Housebound friends, neighbors, or relatives: Offer willing housebound folks plenty of stamps, addressed envelopes, phone numbers, and clear "asks."

Or join them for a movie (see "Host a Living Room Film Fest," page 168), and afterward discuss the film's issues and possible actions.

A copy of this book: Turn to appendix D (page 263) and scan, copy, or tear out Kathleen Dean Moore's letter to her generation of grandparents and send it — with a sweet explanatory note — to your favorite elder.

66. Host a Living Room Film Fest

If you watch a scary movie together, then the scariness is cut in half!
— HIDEKAZ HIMARUYA, *Hetalia: Axis Powers, Vol. 2*

When *Gasland* director Josh Fox brought his film *How to Let Go of the World and Love All the Things Climate Can't Change* to Eugene, hundreds of us packed a local church to watch. As the film ended and credits rolled, Josh broke out his banjo and, with Gabriel Mayers on guitar, got us up dancing in the aisles.

After the revelry, Josh asked, "What'ja think?" Audience members talked until the church closed, and then we reconvened at a pub to sing with the duo and play backup on slide whistles and kazoos from my instrument basket. It was a powerful night of holy-shit education paired with the kind of community bonding that's sealed by listening, singing, and swing dancing together in a church sanctuary.

You, too, can host an eco-film party for your tribe of family and friends — or even for one child — with just a screen and a room. Or, for a larger "festival," create an outdoor cinema: Plug your laptop into a projector and light up any flat wall or hung sheet in your backyard, driveway, or apartment courtyard.

Here's how it works in five easy steps:

1. Rent a pro-planet film. See the suggestions below or visit Common Sense Media (www.commonsensemedia.org/lists/environmental -movies).
2. Spread picnic blankets. Help kids make Crunchy Caramel Corn or Forrest's Too-Spicy-for-Mom Popcorn (recipes below).
3. After the movie, discuss the issues raised. Help kids listen and participate. Ask: What did you learn? What was the most memorable moment? What surprised you? Are you inspired to take action? What kind?
4. Brainstorm ways to manifest these ideas into action. Lay the ground rule: *No criticizing any idea.* Unleash imaginations. Who knows what wonderful alliances or schemes might be hatched in your living room?
5. Afterward, get music going, push furniture aside, and boogie on (see "Kick Out the Climate Blues," page 164).

If you have young kids...

Nature documentaries:

Bears
Born to Be Wild
Cosmos: A Spacetime Odyssey
Earth (Disney's)
Jane
Knut & Friends
The Man Who Planted Trees
March of the Penguins
Winged Migration

Fictional and fictionalized stories:

Avatar
Dolphin Tale
The Lorax
Wall-E
Fly Away Home
FernGully

If you have older kids...

Political documentaries:

A Fierce Green Fire
Food, Inc.
The Great Invisible
In Defense of Food
Plastic Paradise: The Great Pacific Garbage Patch
Taking Root: The Vision of Wangari Maathai
Virunga
We the People 2.0
Who Killed the Electric Car?

Climate documentaries:

Before the Flood
Chasing Ice
Chasing Carol
Gasland

*How to Let Go of the World and Love All the Things Climate Can't
     Change*
· *An Inconvenient Truth*
No Impact Man
This Changes Everything
Years of Living Dangerously (TV series)

If you want to eat...

Crunchy Caramel Corn:

3 tablespoons coconut oil, divided in half
⅓ cup popcorn kernels
¼ cup maple syrup (or honey or agave nectar)
½ teaspoon salt
1 tablespoon vanilla extract

In a large pot with a lid, heat half the coconut oil until it melts, then add the popcorn kernels. Cover, and cook on high until the popcorn is popped. In a saucepan, combine the remaining coconut oil, maple syrup, and salt. Boil 2 minutes, remove from heat, and stir in vanilla. Drizzle over popcorn and toss to coat popcorn evenly. For an extra-crunchy treat, spread caramel popcorn evenly onto two cookie sheets. Bake on 250°F for one hour. Enjoy!

Forrest's Too-Spicy-for-Mom Popcorn:

Cook the popcorn in the same way as the recipe above. Then make the coating:

1 tablespoon butter, melted
1 tablespoon Bragg's Amino Acids or tamari sauce
2 tablespoon brewer's yeast
Cayenne pepper to taste

Mix all ingredients in a bowl. Pour over popcorn and toss around. Use as much cayenne as you can handle. Wash hands (before rubbing eyes). Warn Mom before she eats any. Enjoy!

67. Tame Your Tongue

The goal of any true resistance is to affect outcomes, not just to vent.
— ARIANNA HUFFINGTON

I once took my baby and toddler to a reunion requiring three flights, a boat ride, and a stagger up two flights to a house full of relatives I hadn't seen in years. Lacking my husband's strong arms and steadying presence for the grueling travel, I adopted a mantra: *I request what I need using non-violent communication.*

Nonviolent communication (NVC) improves communication with everyone. Founded by Marshall B. Rosenberg and applied equally well to both personal and international conflicts, NVC is a process that helps people connect with one another's humanity.

The principle is this: *I care about you and seek the same care, so we'll listen to one another and work toward meeting both of our needs.* Apparently, healthy humans who aren't feeling threatened display a basic desire to get along.

Here's the NVC, boiled-to-the-basics script for managing a conflict. Importantly, it ends with a clear "ask," not a list of accusations:

> When _____ happens, I feel _____ because I need to
> (or I value) _____.
> Would you be willing to _____?

Others are free to accept, propose an alternative, or refuse (though refusal can have consequences, such as the other person ending participation in the relationship). This script served me well through my reunion journey — even the relative-refusing-to-smoke-away-from-infants episode — until the final six hours.

As a heat wave baked our crowded gate at Chicago O'Hare, a harried gate agent refused to gate-check my stroller, something agents had done on my previous five flights. I was hungry. Our plane was delayed. My baby cried to nurse and had a loaded diaper. The bathroom was far off and the logistics were daunting. All I could think was, *Must ditch stroller.*

My pleas for compassion and assistance were met with a snarled,

"Take your stroller and step aside, Ma'am!" Instead of following my trusty nonviolent communication script — which urges that when asking nicely fails, you either yield or politely request to speak with the manager — I snapped, turned my back on the agent, and abandoned the empty stroller in a "This is ridiculous!" huff.

Ten steps into my getaway with my kids, a bulky police officer blocked our path.

He boomed, "Ma'am, retrieve your stroller now or you'll face arrest and a thousand-dollar fine." Needless to say, a minute later, I was pushing that stroller.

Why does this matter for the climate?

1. Because we can't snap when it involves governments. Their snaps are always bigger.

2. NVC opens doors, helping with the inevitable conflicts that arise between humans. Government officials — like airline clerks — are just people who crave kindness, too.

3. People seeking change often rant about problems but forget to make reasonable, do-able, solution-oriented requests, such as, "Would you be willing to come to our rally? Would you be willing to contact the governor to ask him or her to deny the coal-mining permit?"

Let's apply NVC to a government climate-change "ask":

When [you threaten to build a fracked gas pipeline], I feel [angry/
    scared/betrayed]
because we need [healthy water, soil, and atmosphere to survive].
Would you be willing to [reject that permit]?

The Case of the Abandoned Stroller resolved when I retrieved it, realizing I'd foolishly escalated a low-stakes situation and invited a forceful response. Avoid this, if at all possible, unless accepting "no" isn't an option (then see "Break Human Laws (to Obey Nature's Laws)," page 234).

If you have...

Five minutes: Learn the NVC basics through the Center for Nonviolent Communication (www.cnvc.org).

One hour: Teach your children to put their needs into the NVC script. Practice together when you have a family conflict, and help them use it for conflicts with a friend, teacher, or government.

And/or testify or write to a local lawmaker about climate change using this script.

Eight hours: Take a local "Introduction to Nonviolent Communication" workshop through the Center for Nonviolent Communication.

68. Let Your Farmer Feed You

I don't grow veggies. I grow children.
— My friend Diane

I can't believe I'm admitting this in a book on climate change, but (shhhhh) *I do not garden.* You do, perhaps. Maybe we all should, but here in gardener's mecca, I can't kneel, thanks to a football injury (okay, I was on the sidelines), nor shovel compost or sledgehammer posts, all central activities to growing food.

But that's not the main reason I don't garden. Food production takes time and effort I'd rather put into more effective climate action. If we truly, madly, down-in-our-bones-ly wish to protect the environment and our kids, then let's honestly assess every activity we consider "climate action." If we must choose between gardening and stopping pipelines, the latter is, right now, the more planet-friendly choice.

Of course, if you want to garden, please do. It's good for the soul (scientists have found dirt makes us happier), and children benefit from it, reveling in strawberries-from-dirt magic and learning from the soreness in their muscles how much work it takes to press apples into cider. If you're happily feeding your family from your backyard, don't stop.

But if you're gardening to save the planet, guess what. It won't. During this holy-shit moment on Earth, it's far more critical to enact bold climate-justice policies than to shrink your family's wee footprint. Rather than grow your own calories, it's better to get yourself — by Hummer, if necessary — to city council meetings and town halls to demand policies that break dirty energy's stranglehold on everything.

The same principle holds for recycling, composting, and any time-intensive household chore like hanging laundry. Yes, it is essential to keep food from the landfill, where it outgases methane and wastes precious nutrients that could amend our soil, but the point is *We need laws* mandating curbside composting in cities and funding for its infrastructure. *We need laws* requiring that the food industry become zero-waste. *We need laws* requiring methane capture at feedlots. *We need laws* revolutionizing agriculture so that carbon is sequestered, not spewed. *We need laws* to end huge agri-business subsidies and instead incentivize small family farms.

We need laws so humanity obeys the laws of physics, which reign supreme no matter how shrilly denialists insist otherwise.

So, to save the planet, let your farmer feed you. Support family-owned farms through community-supported agriculture and shop at co-ops, natural grocers, and farmer's markets, which help keep the food economy local, instead of supporting corporate food giants. If you can afford it, buy organic, since it's an important, system-changing choice.

Let's use what little free time we have to remove people making Stupid laws and install nice people who actually care about kids, food security, and Smart climate policies. When that's done, by all means — let's plant strawberries.

If you have...

No time: Leave your lawn alone. Forget the garden or planting native species (unless you want to). A lawn does sequester some carbon, and using a push mower is eco-friendly (or try electric, which is cleaner than gas, especially when utilities use renewable energy).

An hour: Use LocalHarvest (www.localharvest.org) to find and support your local farmer's market or join a farm through the popular community-supported agriculture (CSA) program. Flexible CSA programs include shares for all kinds of products, like flowers, meats, and dairy.

An afternoon: Search online for a local farm loop or a self-guided tour of farms open to the public. Pick lavender, pet alpacas and sheep, get lost in a corn maze, and take home some local wine. Many farm loops are bicycle-friendly.

Two to four hours a month: Join a food co-op. Work a shift, attend a meeting, befriend fellow members. You'll support local farms and build community.

Support — or start — a farm-to-school program in your school. For inspiration, check out the successful Village Kitchen at Eugene's Village School (http://happyvillage.org/kitchen).

69. Hound Representatives

Unless someone like you cares a whole awful lot, nothing is going to get better. It's not.
— DR. SEUSS, *The Lorax*

I've become something I never wanted to be, something my parents never were nor wanted me to be. I have become a pain in the ass.

I only wish I'd started sooner.

On Mondays, I call Washington, DC, and thank/criticize my member of Congress, a government agency, or a congressional committee head for what they're doing/not doing to keep fossil fuels in the ground. Members of Congress pass laws that either will or won't protect democracy, justice, and the planet, so I make specific demands, such as, "Don't drill the Arctic National Wildlife Refuge!" I make several calls weekly, spending one minute per call.

Before my life as a pain in the ass, I signed countless petitions online — so easy, polite, and civilized! The I-don't-have-to-talk-to-strangers aspect jived with my semi-introverted nature. But when climate deniers took over our government, I knew planet-saving required raising my voice.

Indivisible is an easy-to-read, twenty-six-page document that describes how to do that effectively. Written by former congressional staffers after the 2016 election, it details how the Tea Party, representing a minority of Republicans, has successfully influenced the individual politicians who write our laws, and it suggests we quickly adopt their tactics.

To summarize *Indivisible*:

1. Members of Congress (MoC) don't give a crap about online petitions or emails. Mailed letters — also ignored. What matters? Public humiliation. Anything that tarnishes their reputation as a hardworking MoC heroically defending their constituents' interests. They fear losing votes, so they gauge their popularity by counting calls and appeasing whoever assails them — or thanks them — publicly, loudly, and relentlessly.

2. Face-to-face encounters matter, so attend town halls. Bring people willing — or even eager — to speak in public. Meet early to distribute questions, then fan out around the room. Demand answers. Be polite, but we're battling for planet Earth, so corner your MoC and don't back down.

    Go to your MoC's office and attend their events. One Oregon senator holds monthly coffee chats with Oregonians visiting DC. Go to ribbon-cuttings and baby-kissings to publicly thank — or question — your MoC. Talk to reporters and protest so creatively that your save-the-planet message becomes the story.

3. To really get your MoC's attention, call daily. Senior staff and senators get daily reports of most-called-about topics. Make the planet one of those. Republicans generally bother MoCs *four to eleven times more often than Democrats*. Let's change that.

Here's what to do when calling DC offices:

Ask for the staff member in charge of your issue.

Give your zip code.

Make it personal: "I voted for you last time..." or "As a parent, I'm appalled that..."

Make one or two specific demands, not six.

Be clear in your "ask": "The senator should/should not support... which will win/lose my vote."

Pain-in-the-ass-ness gets easier with practice. If I can do it, so can you. We also have to say a hearty "Thank you!" to representatives fighting for the planet, letting them hear *publicly* — at town hall meetings, in letters to the editor — that we have their backs.

To quote *Indivisible*'s authors: "Get to work. We will win."

If you have...

Three minutes: Join the Indivisible campaign (www.indivisible.org), The 65 (http://thesixtyfive.org), or 5 Calls (https://5calls.org). Follow them on Facebook or Twitter; get email alerts. Using their resources, find the number for one congressional representative and call with your demand. Tie your specific ask to pending votes or events.

An hour: Read the *Indivisible* guide (www.indivisible.org/guide) and sample script. Program numbers for all three of your members of Congress into your phone (including their local and DC offices), and file under *P* (for "politician"), for easy future reference.

Then, read the (fascinating!) leaked Tea Party document detailing their tactics (http://d35brb9zkkbdsd.cloudfront.net/wp-content/uploads /2009/07/townhallactionmemo.pdf). Use them.

An afternoon: Visit a representative's office with a group of constituents. Request a meeting. Protest. Review *Indivisible* for successful strategies. Remember, there's strength — and fun — in numbers.

70. Demand Clean Energy —
While Filling Your Tank

Of course we [use fossil fuels to fight fossil fuels], and people in the North wore clothes made of cotton picked by slaves. But that did not make them hypocrites when they joined the abolition movement. It just meant that they were also part of the slave economy, and they knew it. That is why they acted to change the system, not just their clothes.

— NAOMI ORESKES, Harvard historian

The problem with being a fighter for our kids while also living like most everyone else on the transportation front — driving a gas-powered, climate-killing car — is that I end up regularly drinking from the trough of the fossil-fuel beast I'm trying to slay. It's an odd dependency that oil companies and anyone loathe to change gleefully seize on as evidence of "hypocrisy," but the story isn't that simple.

I'm a captive. I know, sounds victim-y, but it's true. We're all captive to a dumb and dangerous dirty-energy infrastructure.

Of course, some people shrink their personal carbon footprint by, for example, using electric cars, solar panels, and bamboo flooring. Do all of these, if you can. The footprint of an EV is half — or less — of a gas car (see "Keep the Clunker," page 34). Buying EVs and solar panels and eco-friendly bamboo builds demand, which is important when the federal government is passing Stupid, climate-killing laws.

But many families can't afford to install personal green infrastructures. And everyone lives within our society's larger fossil-fuel economy. No matter how hard we fight for systemic change, we all occasionally — if not regularly — hand money to BP, Exxon, Shell, and other corporate monsters funding climate denial and heating the world.

I used to feel ashamed during the fill, like a preacher who rails against the evils of the flesh only to sneak over to the brothel on Saturday nights. I always wanted to leave the scene of the crime quickly.

One day I thought: *What if, instead of feeling ashamed, I talk about this paradox with our kids? What if I did something about it in those few minutes? What if I demanded a fossil-free future every time I got gas?*

So I started making calls at the pump. As I did, I learned that the most effective approach is to join a coordinated campaign, since it floods offices with phone calls at key times as decision makers are considering new laws. I got the 5 Call app on my cell phone, which tells me who to call about that week's most pressing climate legislation. Now I have fun. I put my conversation with elected officials and agency office staffers on speaker phone so my teens see me turning shame into a small measure of empowerment. And I have my first calling "streak"! (A streak counts the number of calls I've made that week.)

Consider this: If even 3.5 percent of us make these calls every time we go to the gas station, decision makers will receive hundreds of thousands of pleas daily for a fossil-free future — from families saying, poignantly or hilariously, "We want to go fossil-free!"

If you have...

One minute: Get on 5 Calls. Go to their website (https://5calls.org); get the free 5 Calls app, sign up for weekly email alerts, or follow #5calls on Twitter. Enter your zip code, and voilà, the office, number, and a sample script appear.

One minute, while pumping gas: Make a call! Start a call streak! It's easy, especially with the script. And politicians' staffers are polite.

Fifteen minutes: With kids, make a car sign, "We'd rather be riding zero-emission buses and trains!" Attach to the car's bumper or post it in a rear window.

An hour: Start a movement. Film yourselves calling your governor from the gas station. Make it a performance, like a singing telegram–style message, with kazoos and fun costumes. Post it on YouTube and invite viewers to join you in gas-station activism by contacting elected representatives while at the pump. Promote 5 Calls and brag about your call streak!

71. Escort Big Oil from the Museum — and Off Campus

We have a moral obligation to get on the right side of the issue of global warming.
— JAMES POWELL, former director of the Natural History Museum of
   Los Angeles County

One of the richest guys on Earth is using the Smithsonian Museum to mislead our children. Billionaire oil baron and antigovernment zealot David Koch paid the museum $15 million for its David H. Koch Hall of Human Origins, which, since 2010, has taught our kids that humans have adapted in the past to "dramatic climate changes" and strongly suggests that, by golly, we're going to do it again!

In one especially misleading section, visitors are told to imagine a far-off future time (not, of course, right now!) when the Earth is "really hot." Two cartoon guys — one with an extra-long torso, the other with especially sweaty armpits — smile benignly from a panel that asks, "How do you think your body will evolve? Will you have a tall, narrow body like a giraffe? Or more sweat glands?" You're invited to vote for your preferred body.

Yes, seriously.

Even if I weren't nine, I might reasonably conclude that global warming's no biggie. If the Earth really does ever, maybe, someday warm up, my body will just "evolve." (I'll take the extra sweat glands, please!)

Here's the thing. When we hear "Smithsonian," we think credible, well-researched, peer-reviewed science. We don't think of directors cutting deals with climate-denying polluters. We don't imagine museums selling propaganda space in *the* most visited museum on Earth.

"When our kids go to museums," my mom-friend Melissa fumed on learning of the exhibit, "they assume that what's on the walls is true. They're in a holy place of learning and can't fathom that it's being used as a tool of the fossil-fuel industry."

Time to fathom it — along with the oil industry's, and particularly the Koch brothers', agenda: the end of government and its regulations, especially for polluters. To this end, the Kochs paid nearly $900 million in the 2016 election to install a climate-denying Congress, which is repaying the favor by gutting civil and environmental protections.

But there's good news: Scientists and activists are fighting back. In 2015, in an unprecedented open letter to science and natural history museums, 150 of the world's top scientists condemned Koch Industries as "one of the greatest contributors to greenhouse gas emissions in the United States." They decried the brothers' funding of a vast network of organizations that deny climate change, and they called on museums to cut all ties with the fossil-fuel industry.

After two hundred thousand of us added our names to the letter, the American Museum of Natural History did indeed cut its ties, and Koch resigned from the board. Six other museums divested or cut ties, and more institutions worldwide are divesting daily. (The Smithsonian, though it usually presents such stellar science, at least at this writing, still stands by its Koch-funded, climate-destruction-whitewashing exhibit.)

Parents can join the worldwide "fossil-free culture" movement by demanding that all educational and cultural institutions stop taking money from coal, oil, and gas companies — and remove industry representatives from their boards.

Remember, the tobacco industry once pulled the same junk-science stunts — that's where oil companies learned the trick — but cigarette advertising and sales are now severely restricted and require health warnings.

Whaddaya say we make sure the coal, oil, and gas companies meet the same fate for their toxic product?

If you have...

Five minutes: Join the Natural History Museum's campaign against fossil-fuel influence in museums (http://thenaturalhistorymuseum.org).

Call the Smithsonian (800-766-2149) to demand it present accurate science — and cut ties with Big Oil.

Visit UnKoch My Campus (http://www.unkochmycampus.org) and find out if your college-age child attends one of the 246 colleges that sell academic influence to the Kochs.

An afternoon: Visit a museum with kids and critique exhibits for accuracy and point of view. If necessary, lodge a complaint over misrepresentations of climate science, and write an opinion piece for your local paper about how the museum misrepresents the reality and causes of — and solutions to — global warming.

72. Blow Kisses to the Brave

"I don't feel very much like Pooh today," said Pooh.
"There there," said Piglet. "I'll bring you tea and honey until you do."
— A. A. MILNE, *Winnie-the-Pooh*

One Sunday morning, my email announced I had a Google alert. Hundreds of people were commenting on my spring 2014 opinion piece in *The Oregonian* entitled "Climate-Conscious Oregonians Must Fight the New Deniers." In the article, I lamented how business-as-usual by industry and government is baking our kids' planet, and I wondered what it would take to spark citizens to climate action — perhaps this soccer mom lighting herself on fire?

Most people understood I'd exaggerated to make my point (see also "Light Yourself on Fire," page 248), and in the anonymous reader comments, several called me "wise" or "funny." Others declared me "insane, cultist, brainwashed, hysterical, deranged," and my favorite, "in need of medication!!" Then, emails arrived accusing me of being deluded about planetary heating. The industry-funded group Climate Depot had published my photo and email on their home page, requesting that people fill my in-box.

My teens laughed and high-fived me. "Good job, Mom! You hit the big time!" But it's really not funny. Online gangs regularly thrash researchers and journalists who publish important information on global warming and its effects, as well as ordinary citizens like me who advocate switching from dirty- to clean-energy sources. Advocates sometimes even receive death threats.

After a famous commentator turned on my friend Dr. Kari Norgaard, author of *Living in Denial: Climate Change, Emotions, and Everyday Life*, his fans dutifully harassed her for a while. Fearful, Kari canceled her next public appearance and kept a low profile.

"I was afraid I could be physically attacked," she told me.

Intimidation works to silence — or at least slow down and deeply discourage — good people from speaking up for the planet. Virtually anyone who publicly suggests we stop the fossil-fuel industry's energy stranglehold gets their butt kicked online, often with the help of industry-funded trolls.

What can parents do? Love up our climate defenders who speak publicly. They're vulnerable. They need to know we've got their backs. Expressing gratitude and offering encouragement won't put you in the spotlight, and it can encourage those who are being bullied online.

"Supportive notes made all the difference when I was afraid," Kari said. "One friend even offered to guard my house." Touched, she and her son later sent encouragement to others in the same position. They drew handmade thank-you cards for organizers working to end the buffalo slaughter outside Yellowstone Park.

"I wanted to support others as I was supported," she says. "And it was a relaxing way to connect with my son."

So go ahead, give a stranger a quick "hang in there" email. Fire off a short letter to the editor supporting the publication of science-based climate articles. You'll be like Samwise Gamgee in the *Lord of the Rings*, Frodo's faithful companion to Mordor. Or the hundreds of people who send Katniss that special Mockingjay kiss in *The Hunger Games*. You never know — to a climate hero who feels alone or afraid, your gesture could make a world of difference.

If you have . . .

Five minutes: Thank someone for their positive climate action, either online or via email. Thank students who deliver a speech, display a poster, or read an essay calling for climate action. Remember, everyone, especially elected officials, loves kudos. Repost good climate articles reminding readers such writing takes courage. (Note: Don't argue with trolls or commenters. Post what you have to say and don't read further.)

One hour: Help kids draw a handmade thank-you card to anyone they admire, then mail it, or scan and send by email. For inspiration, check out DoSomething.org's campaign to thank park rangers (https://www.dosomething.org/us/campaigns/wildlife-cards).

Write a note yourself, remembering teachers who teach accurate climate science, sometimes risking administrative or public hostility.

Ninety-six minutes: Watch *Merchants of Doubt* with your kids (age twelve and up). This documentary exposes the industry-hired flacks who dog climate scientists and spread confusion about global warming.

73. Become a Victory Speaker

[A Victory Speaker] need not be his [or her] community's wealthiest or most prominent man or woman, but — banker or carpenter, salesman or clergyman, housewife or school teacher — he or she must be an individual whose character is outstanding.

— WORLD WAR II *SPEAKERS' BUREAU MANUAL*

After the Pearl Harbor bombing, our country transformed itself almost overnight. Declaring that the war must be won *no matter the cost*, President Franklin Delano Roosevelt ignited the engines of government and industry to build up the military. He and Congress raised taxes, passed laws like the "Act to Promote the Defense of the United States," and gave companies juicy contracts guaranteeing profits if they made the machinery for war. By 1942, industries hummed along in a government-coordinated symphony of production fueled by fear of our shared enemy.

Meanwhile, men left to fight and women took over factory work. Families — doing whatever they could to win victory and bring fathers and brothers home — recycled everything, stopped pleasure driving, planted "victory gardens," bought war bonds, and endured rations on meat, butter, sugar, coffee, gas, and shoes. Kids helped, too, collecting tin scraps — toddlers even picked up gum wrappers — to make airplanes.

You know where I'm going with this, right? Ordinary people helped win a world war. We can do the same with climate! These days, we don't need warships but wind turbines, solar everything, powerful batteries, electric buses, biogas digesters, and much more. As before, we must conserve, reduce, and reuse — not to bring down the Axis powers, but the carbon count.

Banding together against a common threat requires courage, inspiration, and clear direction, and FDR provided that during his "fireside chats" — informal radio talks meant to mobilize citizens for the war effort. But inspiration also came from victory speakers — ordinary citizens who volunteered to give daily four-minute pep talks in their community. They appeared at bridge clubs, PTA meetings, or before *Gone with the Wind* played at the local theater. Topics included "What Is Good

Morale?" "Traffic Control During Blackouts," and "Why Save Sugar?" Ordinary people heard from community members they trusted. Inspired, nearly everyone took action.

Unfortunately, in today's war against greenhouse gases, we have no FDR offering inspiration as we gather around the radio. We have no Winston Churchill reminding us that, "without victory, there is no survival." Our own government isn't even pretending to be on our side.

That's why we need victory speakers more than ever — ordinary people who inspire others to join climate recovery efforts. In the internet age, even introverts who shy from the spotlight can create a victory "speech." All that's required is a little creativity and a camera.

There's a side benefit: When our kids see us making victory speeches, they understand, in a deep way, "Dad's not just lecturing me. He's telling *other people* something important." Nothing conveys our values to our kids more powerfully than taking positive, life-affirming action, especially if it's challenging.

And if our government continues to betray everything for which citizens sacrificed in World War II, our children will benefit from seeing us publicly model a new form of resistance.

If you have...

Twenty minutes: Read Bill McKibben's *New Republic* article "A World at War" (August 15, 2016), about solutions to accelerate our clean-energy transformation (https://newrepublic.com/article/135684/declare-war-climate-change-mobilize-wwii). The whole thing's a victory speech!

Forty-five minutes: Make a friendly, legible sign that invites people to take one action for your local climate fight: "Reject the oil terminal permit!" "Jane Smith for school superintendent! She supports climate literacy!" With kids, hold it on a street corner for twenty minutes. Wave. Smile.

A weekend: Get trained at Al Gore's Climate Reality Leadership Corps (https://www.climaterealityproject.org/leadership-corps).

Bored kids: Challenge kids to make a thirty-second video "victory speech." Have fun with it, upload it to YouTube, and send links to everyone.

Try a twist on the lemonade stand: a climate-aid stand to help fund a local climate campaign (free lemonade for donations over three dollars). On a laptop, play kids' victory speeches!

74. Trumpet the Pope

What kind of world do we want to leave...to children who are now growing up?
The Earth, our home, is beginning to look more and more like an immense pile
of filth.

— POPE FRANCIS

In his 2015 papal letter addressed to "every person living on this planet,"
Pope Francis — considered the most-respected leader on the planet —
issued a thundering call for a "bold cultural revolution."

"Doomsday predictions can no longer be met with irony or disdain,"
he said, and he went on to scold "obstructionist" global-warming deniers
"concerned with masking the problems or concealing their symptoms."
He even chastised politicians who cater to the fossil-fuel industry.

Then, Pope Francis flew to Washington, DC, to school those very
politicians, the Republican-dominated US legislature, on the eve of Paris
climate talks. He challenged them boldly to "do unto others as you would
have them do unto you" in regard to refugees, the poor, and care for the
Earth.

Many Republicans mocked him, but the majority of the world
embraced his message. For religious leaders worldwide — and for the
1.2 billion members of the Catholic Church — it was a sea change.

For the first time ever, the powerful Vatican had explicitly framed
global warming as a moral issue.

Within weeks, 1.8 million people signed several faith-based petitions,
and people around the globe mobilized for more than two thousand vibrant protests to pressure leaders to hammer out a meaningful climate
agreement.

Since then, other faith leaders have amplified these calls for climate
justice. The Dalai Lama called global warming a "problem which human
beings created," and he said, "All of humanity [is] now responsible for
taking action."

Archbishop Desmond Tutu went further. He said, "Time is running
out.... We can no longer continue feeding our addiction to fossil fuels....

We must support our leaders to make the correct, moral choices." He then issued his "Call to Action," including that we demand that our leaders:

1. Freeze further exploration for new fossil sources.
2. Hold those responsible for climate damages accountable.
3. Encourage governments to stop accepting funding from the fossil-fuel industry.
4. Divest from fossil fuels and invest in a clean energy future.

All this leadership gives parents a springboard for launching healthy climate conversations, both with our children and with the world at large. Even if people don't accept the reality of human-caused global warming, they have an opportunity to show compassion to those who suffer, innocently, the merciless lash of heat waves, storms, and relentless drought.

Compassion is a powerful angle. People of faith make up the majority of the world's population, and in the United States, 80 percent identify as Christian. At the heart of Christianity is Jesus's love and compassion for the poor, vulnerable, and suffering, which includes those brutalized by climate change.

We can post pro-environmental scripture in our places of worship, which is what eighty-three thousand young activists did for President Trump's inauguration. Linked through the youth activism site Do Something.org, young people copied lines promoting care of the Earth from religious texts and then decorated signs and posted them on bathroom mirrors and walls.

What did those rascally young people call their campaign? Nature's Calling, of course.

If you have...

Ten minutes: Visit the websites of various religious denominations that are taking climate action and see how you might join in:

Buddhist Climate Action Network: http://globalbcan.org
Catholic Climate Covenant: www.catholicclimatecovenant.org
Global Catholic Climate Movement: http://catholicclimatemovement
.global
Jewish Climate Action Network: www.jewishclimate.org
Operation Noah: http://operationnoah.org
Young Evangelicals for Climate Action: www.yecaction.org

One hour: Help kids launch a Nature's Calling campaign in your family's place of worship, perhaps with support from DoSomething.org (www.do something.org/us/campaigns/natures-calling).

Ask religious leaders to advocate climate justice in a sermon, public letter, or by wearing their religious vestments at a public rally. Arrange for a guest sermon from a religious climate-action leader at your place of worship.

Time to read: Read the pope's encyclical on the environment, *Laudato Si* (https://laudatosi.com), which is free online. Also read *Moral Ground: Ethical Action for a Planet in Peril*, edited by Kathleen Dean Moore and Michael P. Nelson.

75. Shout from the Solar Roller Coaster

You never change things by fighting the existing reality. To change something, build a new model that makes the existing model obsolete.
— R. BUCKMINSTER FULLER

Nature designed us to obsess on the negative. Tell me ten things I'm doing well and one that "needs work," and I'll forget the ten compliments and fixate on your single criticism, even processing it in a deeper-thinking part of my brain. Ask about my childhood, and I'll tell stories of broken bones and teenage heartbreak, forgetting the good. You'll do the same.

Right now on the climate front, we're hammered with bad news, both about its terrifying impacts and regarding the foolish decisions of those in power. The ratio seems reversed: For every positive development, there are ten negative ones.

This is why we've got to celebrate every win, even small ones. If we can pause and resist our natural impulse to dismiss victories and mourn losses, it'll help lift our spirits and remind us that the clean-energy transformation is already happening. And can be a lot of fun.

Given where things stand right now, I consider it my sacred duty to shout good news often and exuberantly. Parents can do that in many ways, including through mealtime, and sideline, conversations, social media, and letters to the editor. Since editorial-page rants and finger-wagging rarely inspire people, why not instead effervesce about local schoolkids learning to compost, or how American solar employs more people than the coal, oil, and gas industries *combined*?

Better yet, shape the news. Did you know that anyone can send a press release to a local newspaper, magazine, and radio or television station? When my family condemned our neighbors' homes and installed a block-long fake pipeline to protest the proposed pipeline through Oregon, our press releases sent to local editors paid off in front-page news coverage.

You or your young in-house reporters could give a shout-out about forward-thinking farmers, highlighting those who are switching from conventional to no-till farming to improve the soil and capture carbon, or promote a local program offering rebates for new electric cars. We need climate heroes who install solar panels in refugee camps or who sue polluters, and we need heroes who publicize this good work and help change hearts and minds — and inspire hope.

Here's a sample of the kind of good news I shared with my community on social media and by word of mouth in fall 2017:

1. Denmark's biggest energy company abandoned fossil fuels — and became the world's largest offshore wind farm company! (October 2017)
2. TransCanada canceled a major tar sands pipeline project! (October 2017)
3. China is building the world's first smog-eating "Forest City!" — and the first floating solar-power farm! (July 2017)
4. France banned new oil and gas exploration! (June 2017) *Viva la France!* (October 2017)
5. Copenhagen has more bikes than cars! (March 2017)
6. Nevada and Washington passed laws that will create thousands of solar jobs! (July 2017)
7. In New Zealand and Ecuador, rivers now have legal rights! (June 2017)
8. And, since I mention roller coasters: California's Great America amusement park is powered by *100 percent renewable energy*!

Your turn. Find soul-feeding, foot-stomping climate news that uplifts, inspires, and entertains. Then tell everybody.

If you have...

Five minutes: Visit websites that offer good news: 10:10 (https://1010uk .org), CleanTechnica (https://cleantechnica.com), *Yes! Magazine* (www .yesmagazine.org), and the wonderful *Brains On!* (www.brainson.org), which features science by and for kids.

Put something inspiring in your email signature or your voicemail — "Sweden's gone fossil-free. So can we!"

Thirty minutes: Write a press release and send it to the appropriate media editor. For an example, see appendix C (page 260).

Two hours: With kids, craft a peppy letter to the editor celebrating good news about renewable energy.

Watch the humorous 2015 documentary *Where to Invade Next*, in which filmmaker Michael Moore steals the best social innovations from other countries, including six-week paid vacations and Finnish schools with no homework.

76. Push Your City to Get Ready

The bottom line is it's going to be bad everywhere. It's a matter of who gets organized around this.

— BRUCE RIORDAN, director of the Climate Readiness Institute

In late summer 2017, after a hotter-than-ever summer, the Oregon wilderness went up in smoke.

The Willamette Valley was choked with so much smoke that we could look directly at the orange-red sun without squinting. Life in town became eerie as, week after week, we inhaled our beloved forests. Many days, the smoke was so bad that children stayed inside, coaches canceled practices, and those venturing outside for more than a dash to their cars wore masks. And it was *hot*. The only escape was indoors with air conditioners that, ironically, exacerbate global warming. Meanwhile, hurricanes ravaged Puerto Rico and submerged Houston, which destroyed thousands of homes and set a chemical plant afire. A month later, an algae bloom covered much of Lake Erie with green, toxic slime at the same time that blazes swept California, incinerating almost nine thousand homes and killing forty-three people. Friends, depressed and frightened, spoke of apocalypse.

If this is a 2017 snapshot of life on a planet wracked with fever, what's in store for families in, say, the next decade?

In North America, we know to expect more weather that seesaws between drought and deluge. We know we'll have to guard our kids against more disease-carrying ticks and mosquitoes and pay more for groceries as crops are hit. And we know much depends on what we do today to stop Stupid and start Smart.

But much also depends on where we live. The South is scheduled to bake, the West to burn, the Southeast to endure storm lashings, and coastal cities like New York, San Francisco, Seattle, Boston, Miami, New Orleans, and Honolulu to regularly flood and become partly submerged.

Here's the thing, though: Some cities are preparing for this. Most aren't. You might want to be in a city that is — or demand that the one you live in make climate-action and resiliency plans *now*.

Take Denver, for example. The Mile-High City is working to become a healthy place where happy people eat local food and reduce their water and fossil-fuel usage. It's working so that people in "transit deserts" have driving buddies, everyone breathes clean air, and kids can swim and fish in clean rivers.

Or Portland, which has invested heavily in light rail, bicycle infrastructure, LEED-certified buildings, and lowering emissions — and teaches climate literacy to its students. Or Chicago, helping neighborhoods build the community connections that keep people less isolated — and therefore safer — for coming heat waves.

Does your town have a climate-action plan? If not, tell officials to get going. They don't have to reinvent the wheel — just copy cities already doing it beautifully.

If you have...

Eight minutes: With kids, visit the Sea Level Rise Viewer (https://coast .noaa.gov/slr), an interactive website created by the National Oceanic and Atmospheric Administration. Zoom in on satellite photographs of anywhere in the United States to see how the land will change with rising seas. It's sobering, and a bit addictive. My family looked at my old neighborhood in New York's East Village (underwater) and everywhere we have family. What will happen where your family lives? Note that this is the kind of critical information that climate-denying GOP leaders undermine when they slash science-agency budgets.

Twenty minutes: Read the *Atlantic* article "The American South Will Bear the Worst of Climate Change's Costs" (see endnotes), which describes a study detailing the various economic impacts on every US county. How will your region fare?

One to three hours: With older kids, watch a six-minute video about Portland, Oregon's, Climate Action Plan (www.portlandoregon.gov/bps /64076), and read the 2015 Climate Action Plan itself (www.portland oregon.gov/bps/article/531984). See how Smart government policies improve family life in a changing climate and build prosperous, inclusive economies. Demand similar plans for your city, and send Portland's plan to your mayor or city manager. Help kids write letters to the local paper to demand the creation of local climate-action plans.

77. Connect Your Work World to the Real World

Climate change is one of the gravest crises the planet has ever faced. So we took a stand here today at the Bristol County District Attorney's Office.
— SAM SUTTER

Bristol County attorney general Sam Sutter found himself in a bind in September 2014. The father of three was slated to prosecute Ken Ward, also a dad, and Jay O'Hara for blocking a forty-thousand-ton shipment of coal to New England's largest power station. The problem was that Sutter agreed with the activists' banner proclaiming "Coal Is Stupid."

Sutter's not the only parent to face a disconnect between the work world — which feeds our kids — and the larger natural world we must keep healthy to support our kids' lives. Many of us act a bit like Clark Kent, who spends his off-hours as a hero and his workday docile as a sheep. We nag our kids to switch off lights to reduce the family's carbon footprint at home, then turn into Bigfoot at the office, choosing efficiency or profit over protection of our kids' world.

How can parents bring climate heroism to the workplace?

Sutter's solution was so unexpected that it made the front page of the *New York Times*. On what was supposed to be day one of the climate activists' trial, he refused to prosecute and instead marched out to meet reporters and the defendants' sympathizers. Holding aloft a copy of Bill McKibben's *Rolling Stone* climate article entitled "A Call to Arms," Sutter said that the decision not to prosecute was made to avoid the cost to taxpayers, "but was also made with our concerns for their children, and the children of Bristol County and beyond in mind."

He later told *Democracy Now!*: "To the extent that I have a forum, I'm going to speak out about this. And to the extent that my office can be a leader for state agencies, district attorneys on this issue, that's what I'm going to try to do." Then, weeks later, he joined the 2014 People's Climate March in New York City.

Few parents are in a position to splash powerful climate declarations in the *New York Times*, but we can start where we are.

If you want to...

Work collectively: In your company, suggest employers and coworkers do one or more of the following:

Propose that the company divest from fossil fuels and invest in clean energy.

Offer a climate justice scholarship to college-bound seniors.

Model your business after the clothing company Patagonia's financial support for "systemic change" and grassroots activism (www .patagonia.com/climate-change.html).

Organize a service project — like tree planting or canvassing in support of a local candidate who fights for Smart climate policies.

Speak at your city council in favor of local policies that promote clean and renewable energy.

Sponsor a youth climate club the way businesses sponsor kids' sport teams. (Buy them a megaphone!)

Work individually: Consider your profession and the skills and platform you already possess, and work from there. For instance, if you're...

a math or science teacher, then give students problems related to climate math and science.

a farmer, then transition to no-till farming or carbon farming.

a medical provider, then speak publicly about the human health risks climate change poses.

an artist, then help a grassroots group fancy up their website or make protest art.

a lawyer, then offer pro bono services to activists or be a legal observer at protests involving nonviolent civil disobedience.

an investor, then steer clients from dirty- to clean-energy investments. Offer a free public workshop on how to divest and still protect the family nest egg.

a music teacher, then help students write or perform songs about climate solutions.

a marketer, then help your local climate group improve its public outreach.

a real estate agent, then help people value homes threatened with eminent domain on pipeline routes or with sea level rise.

a writer, then help convey the climate narrative, especially to urban populations.

a butcher, baker, or veggie-quiche-maker, then feed volunteers.

78. Know Exactly What You're Voting For

There's no difference between Democrats and Republicans.
— Oft-heard comment during presidential elections

Maybe you've felt or said it. Even my mostly progressive friend Fred says it now and then.

Voting for either party can seem meaningless, and people have grown increasingly disenchanted with our electoral system. The 2016 presidential election made painfully clear the systemic failures of our democracy and the impact of corporate money on elections. I join the majority of both Democrats and Republicans calling for electoral reform.

However, in a book for parents about our climate revolution, I'd be remiss if I didn't make one thing crystal clear:

Right now, on climate policy alone, the Democratic and Republican Party platforms are polar opposites.

Once upon a time, Republicans supported conservation. President Nixon signed the Clean Air Act and emphasized the importance of "clean air, clean water, and open spaces for the future generations of America." That was a half century ago.

Today, a handful of Republicans support pro-environmental policies, but voters must clearly understand that today's Republican Party and its platform *do not*.

Here, take a look at this quick comparison:

Democratic Leadership in 2017	Republican Leadership in 2017
Calls climate change an "urgent threat" and proposes cutting carbon emissions by 80 percent.	Calls climate change a "hoax."
Unveils plans to become "the clean energy superpower of the 21st century."	Works to revive the coal industry and increase offshore oil drilling. In 2016, appointed CEO of ExxonMobil to head State Department.

Supports the international Paris agreement.	Withdrew United States from Paris agreement, isolating us on the world stage.
Supports the Environmental Protection Agency (EPA) and regulations to reduce greenhouse gases.	Calls EPA regulations a "disaster," chose climate denier to head agency, and has proposed slashing EPA's core budget roughly in half, threatening to undermine enforcement of laws protecting our local water and air. Wants agency and regulations gone.
Opposes the Keystone XL pipeline.	Permits Keystone pipeline expansion.
Supports future federal decisions that "contribute to solving, not significantly exacerbating climate change."	Expedites Dakota Access Pipeline, despite bitter opposition from indigenous groups.
Supports a price on carbon.	Opposes a price on carbon.
Supports the Clean Power Plan.	Killed the Clean Power Plan.
Supports prioritizing renewable energy over natural gas.	Cut energy programs to help us expand rooftop solar panels, electric-vehicle batteries, LED lighting, and more. Has eliminated international climate change programs, research programs, and the coal mining health advisory panel.

To be even clearer still: The climate platform and voting record of Democrats deserves criticism, too. Democrats are not doing nearly enough, and climate scientists insist we must strengthen, expand, and build on even the best Democratic proposals. Yet Democrats want the United States to help — and lead — worldwide efforts to draw down greenhouse gases.

Meanwhile, Republican lawmakers are working feverishly to block those efforts.

Now more than ever, every single election makes a difference. So when we approach each ballot, we should know very clearly: Which candidate will pass laws to combat global warming? Which will throw gasoline on the problem and fan the flames?

In the 2016 election, my friend Fred was not going to vote. But when I shared the important differences on climate policies between the two major parties, Fred changed his mind — and voted for candidates fighting for Smart climate policies.

You can do the same with your friends, family, and children, who will soon vote themselves. Combat cynicism, change minds, and — of course — vote.

If you have...

Fifteen minutes: With older kids, read aloud and compare the energy and environment sections of each party's 2016 platform, which can be found on their websites (www.gop.com/platform, www.democrats.org /party-platform). The GOP platform is an especially alarming — and therefore educational — document that accuses Democrats of being "extremists" trying to "sustain the illusion of an environmental crisis."

Read the *National Geographic* article "A Running List of How Trump Is Changing the Environment" (see endnotes), which offers concrete examples of the real-world changes being wrought by a Republican administration hostile to the environment.

An upcoming election: Talk to others about the differences between each party's, and each candidate's, positions on environmental protections. If people don't see its crucial importance right now, discuss how it intersects with social justice (See "Link Eco-Justice to All Justice," page 238).

Time to volunteer: Mobilize voters, especially in swing states, to register and actually vote. Join Rock the Vote (www.rockthevote.com), and fight to get money out of politics and to open up the entrenched two-party system for more voices.

79. Bury Your Neighbor's Chicken

If everyone helps to hold up the sky, then one person does not become tired.
— African proverb

One hot summer's day, my neighbor Jen opened my studio door. "Cow is dead," she announced.

Cow, named for her black-and-white plumage, was a favorite in our shared flock. Now, the beloved chicken was lying claws-up in the backyard, near where the children were playing. Cow needed disposal, if only so the kids wouldn't accuse us of traumatizing the other hens, who were forced to step around their fallen sister.

But I had a pipeline to assail. Only ninety minutes remained before the deadline for my newspaper commentary. Yes, the chicken was rapidly decaying in ninety-seven-degree heat, and proper sanitation, at least, required that I grab a shovel.

I kept typing.

A few minutes later, Jen called through my studio window, "My climate activism today is to bury the dead chicken so *you* can finish writing your damned commentary!"

This episode drove home an important lesson. Jen is a busy parent of three who doesn't identify as an "activist," but in her own ways, she's happy to aid my planet-saving mission. She doesn't organize rallies or challenge senators on the editorial page, but her everyday kindness helps *me* do those things.

This is not a new idea. Jacqueline Woodson tells a story, in her beautiful free-verse memoir *Brown Girl Dreaming*, about "Miss Bell," whose white employer threatened to fire her if she saw Miss Bell marching for civil rights. So, Miss Bell didn't march for civil rights. Instead, she hosted secret meetings and fed scores of people who marched over and over.

During the civil rights movement, many people like Miss Bell did the critical work of caring for marchers on the front lines, just as many people worldwide help today by supporting activists in our battle for climate justice. Remember, there are countless roles — big and small, showy and quiet — in our shared planet-saving quest. Sometimes everyday heroism is overlooked. The smaller stories of love, bravery, and sacrifice don't always seem to measure up to more dramatic historic moments. But it's still heroism, and the great actions we spotlight need that background support.

In addition to dead chicken burial, I've gratefully received countless gestures of support, from pots of soup to writing retreats to an occasional wad of bills. I'm learning to not only receive help but ask for it.

I also invite acquaintances to perform micro-tasks. I shop with two lists, a paper list for groceries and a mental list of that week's top climate tasks. If I run into someone who thanks me for "doing so much," I invite them to take one small, time-sensitive action. One week, I requested someone send "Stop Keystone!" notes to the president. Another week, I invited friends to attend a local rally.

People appreciate bite-size invitations because they can engage without committing to a group or meeting. If you're actively engaged, look for ways others can easily help you, then go ahead and ask, whether it's making a meal, attending a screening, donating old sheets (for banner-making), or Twitter-bombing the governor.

On the other hand, if you're a "civilian," look for ways to support your local climate heroes. Trust me, they need your help.

If you could...

Help a hardworking climate hero: List ten ways to offer support. Some ideas: deliver dinner; take their kids bowling; buy them massage gift certificates; listen when they need an ear; weed their garden; fix their bike; offer haircuts or bail money.

Then ask the person what would help the most or choose the ones that appeal to you.

Use help with your climate heroism: List five easy ways friends and family could help you or your family in the next month. Send around an email.

Or print a short menu of easy actions anyone can take. Include pertinent phone numbers, emails, and links. Store it in your purse, bike bag, or glove box, and offer one whenever people thank you for your hard work or express despair about the environment.

80. Rescue Food

Food is meant to be eaten. Treat it with respect.
— SayNoToFoodWaste.org

This week, I learned that one of the most important things we humans can do to reverse climate destruction is stop wasting food. This surprised me until I dug deeper and learned that *one-third* of all food produced worldwide is thrown out yearly. In the United States, it's 40 percent.

On the one hand, this is a hair-raising statistic, given the massive amount of fossil fuels needlessly used to plant, harvest, process, refrigerate, transport, package, prepare, and display all that ultimately trashed food.

On the other hand, it offers a golden opportunity to cut down on 8 percent of the world's human-caused emissions. And we don't need a magic wand or even new technology — just education, new habits, and Smart laws.

First, though, where does all this waste happen? One-third is created at food processors, retailers, and farms. Entire crops often rot in the field because it's unprofitable to harvest them, and carrots routinely get rejected if they're not straight enough to easily pack into stackable bags. Everywhere along the food-supply chain, workers swoop in to remove delicious, nutritious food with spots or shapes shoppers may reject.

But the bulk of food waste — two-thirds, to be precise — happens in households like yours and mine. It's not because we're laggards who let our fridge contents fester (though it *is* true that we're far less skilled at kitchen efficiency than our counterparts in the developing world). The real culprit is food date labeling.

It turns out that those terms "sell by," "use by," and "expires on" confuse us and drive us to trash delicious, nutritious food. This is because, contrary to popular belief, there's no standardized food-date-labeling system. Those dates stamped on milk cartons don't tell us when the milk becomes unsafe to drink. They have nothing to do with safety, actually. They're arbitrarily thrown on by manufacturers who guesstimate when their product will *taste* the best. In fact, a 2016 study on food date labels concluded, "With only a few exceptions, the majority of food products remain wholesome and safe to eat long past their expiration dates."

A first, easy step climate-conscious parents can take to fight home food waste is to ignore date labels and do what our ancestors did for millennia: Trust our noses and taste buds to alert us when food spoils. Second, we can be mindful about not overbuying (which saves money!).

Third, we can demand laws that standardize food date labeling, helping us distinguish between labels that indicate peak quality and those that indicate when items may become unsafe to consume. We can also demand that food be allowed to be sold or donated when it's still perfectly good, but past its quality date.

That has another powerful benefit: matching safe, nutritious food with people who are hungry.

If you have...

Five minutes: Check out Eat By Date (www.eatbydate.com), which aims to "educate consumers on how long food really lasts past its printed date."

On your next shop, grab some clear containers for leftovers. If possible, use glass. (Plastic containers leach chemicals that are especially toxic to children and pregnant women into whatever food they contain.) We also use pint-size canning jars, which are affordable, stackable, and have small fridge footprints.

Thirty minutes: With your kids, read the book *Hungry Planet: What the World Eats* by Peter Menzel and Faith D'Aluisio, who photographed families worldwide at home with their week's supply of food. Ask kids: What would ours look like?

A desire to volunteer: Seek out local food-rescue groups, which ferry edible food to soup kitchens and food pantries. If possible, volunteer with your kids, or suggest teens volunteer with friends.

Kids who like to cook: Challenge them to make a meal from what's on hand in your kitchen — without a trip to the store. Help them adapt recipes, which is an empowering skill for kids to learn so they can run their own efficient, low-waste kitchens in the future.

81. Kick Compost to the Curb

Throwing food into the landfill is insanity.
— BRENDA PLATT, Institute for Local Self-Reliance

B usted. My sister-in-law, Brenda, caught me sneaking little strawberry tops into the garbage.

"Could you pick them out, please?" she asked. "I'll start a compost bin." Our families were sharing a weekend beach rental, and I wasn't a vacation composter unless there was an onsite bin. And I rarely transported rotting food across state lines because, well, parents need vacations, too.

But my brother married a crusader. My siblings and I still have cloth napkins from their unique, zero-waste wedding, and as codirector of the DC-based Institute for Local Self-Reliance, Brenda battles garbage incinerators the world over. She authors reports with titles like "Stop Trashing the Climate," which link rotting landfill garbage to global warming. She promotes waste reduction and composting on any scale.

Even for a few strawberry tops. When I hesitated, tired after being awoken at 2 AM by the teen party — and facing the dirty dishes they had left — she argued, "Mary, composting is the number-one thing you can do to stop climate change!" She picked through garbage. I washed dishes. We laughed about it later.

Soon after, I attended her keynote speech to Northwest waste specialists and finally understood. When we landfill food, she explained, we discard good nutrients that could enrich our soils, which are being rapidly depleted. Worse, that rotting food off-gases methane — the same dangerous stuff leaked by natural gas drilling, fracking, and burping cows. During its first twenty years in the atmosphere, methane holds eighty-six times more heat than carbon.

North Americans toss tons of food into landfills — 21.5 million tons yearly. Some of that waste can be avoided (see "Rescue Food," page 203), and the rest can be composted, which helps communities revitalize soil, improve plant growth, irrigate less, reduce methane emissions, replace fertilizers and pesticides, and create local green jobs. Composted food and yard waste turns into natural fertilizer — nutrient-rich dirt — that farms actually buy. This points to another key cobenefit of composting: It creates twice as many jobs as landfilling, and four times as many as incinerating.

Moreover, when compost is spread on the ground, it pulls carbon out of the atmosphere, effectively bringing it home to the Earth through photosynthesis. It does this so efficiently, in fact, that some advocates estimate that by spreading a mere half inch of compost on 2.7 billion hectares of the world's grasslands, we'd reduce carbon in the atmosphere to the safe concentration of 350 parts per million.

I could almost hear choirs of angels as she spoke. Why isn't everyone joining the revolution to turn spoils into soils?

Happily, several cities and states are. San Francisco, Portland, and Seattle have mandatory curbside composting. Connecticut, Rhode Island, Vermont, and Massachusetts are phasing out landfilling organic waste. New York City has the nation's largest curbside compost program, and intends to make curbside compost available to every resident in 2018.

On the home front, families can compost in a backyard pile or — for those without backyards — with a bin of dirt with tiny worms under the sink. That's an especially fun introduction to composting, says Brenda, because "kids love worms!"

If you want to ...

Start composting: Read Stu Campbell's *Let It Rot!: The Gardener's Guide to Composting*. Watch the four-minute how-to video "Compost Kids" (https://www.youtube.com/watch?v=Njbn34JrKnE).

For worm bins, read Mary Appelhof's *Worms Eat My Garbage!*, or visit the University of Maine Cooperative Extension's primer, "Worm Composting" (www.youtube.com/watch?v=jJ3QIZMta98).

Learn about carbon farming: Visit the Marin Carbon Project (www.marincarbonproject.org/home), or read Kristin Ohlson's *The Soil Will Save Us: How Scientists, Farmers, and Foodies Are Healing the Soil to Save the Planet*.

Advocate for large-scale composting: Tell grocers and restaurants to go zero-waste. Ask school, city, and state officials to pursue a zero-waste path — and to offer curbside compost. If your town offers curbside composting, publicly praise and use those programs.

Learn about the Institute for Local Self-Reliance's Waste to Wealth programs that help communities implement Smart waste policies (https://ilsr.org/initiatives/waste-to-wealth).

82. Do the New Math

The more carefully you do the math, the more thoroughly you realize that this is, at bottom, a moral issue; we have met the enemy, and they is Shell.

— BILL MCKIBBEN

I failed Algebra 2 in eleventh grade. The *entire* year. I haven't done math since, unless you count calculating restaurant tips and avoiding home foreclosure. So it's kind of funny that I married a math teacher and strange to find myself cheerleading a high schooler through trigonometry. And, surprise, I'm going to ask you to do some math right now. Because there's one equation that clarifies this moment of human history, and every family needs to know it.

Ready? We'll do it together:

1. The 2015 Paris climate accord pledged to keep global warming to 2 degrees Celsius tops. (It should really be 1.5 degrees or less, but for this exercise, we'll stick with the agreed-upon 2 degrees.)

2. Humanity has already put enough carbon in the atmosphere to raise Earth's temperature .08 degrees Celsius. To keep heating under that 2-degree mark, scientists estimate that we could, at the very most, *maybe* burn another 565 gigatons before we irrevocably alter the Earth.

3. The fossil-fuel industry has at least 2,795 gigatons worth of carbon in its proven reserves of coal, oil, and gas underground.

4. Now: Calculate which is more — the amount we might get away with burning, 565, or the amount industry plans to burn, 2,795?

If you answered that 2,795 is a bigger number than 565, you're correct. In fact, 2,795 is about five times 565, the amount we can safely burn.

That math is straightforward: Coal, oil, and gas companies plan to dig up, sell, and burn reserves that, by themselves, will pretty much render our kids' Earth uninhabitable. The only shot we have at avoiding this hellish scenario is to keep those reserves underground.

This is, of course, an existential threat to Big Oil, which points to those underground reserves and tells investors, "We're going to dig those up and sell them! Give us money to do that!" This prompts investors to write massive checks. That's why industry denies the above-mentioned math, desperately trying to reassure its investors that, no, all is well! Really!

To raise the alarm that dirty-energy companies plan to incinerate the planet, 350.org cofounder Bill McKibben and several young leaders of 350.org toured campuses all over the United States for their 2012 Do the Math tour. McKibben told the huge crowds: "It's a lot like nuclear overkill: We've got five times the carbon that we need to cook the planet already in our arsenal. But this time we're clearly planning to go ahead and push the button."

Everywhere they went, they challenged audiences to join them in confronting the fossil-fuel industry head-on. And everywhere they went, they left youth-led divestment campaigns — 252 by the tour's end — in their wake, organized by determined students the *New York Times* called the "vanguard of a new national movement." In the chapter "Divest. Get Everyone To." (page 209), we'll learn how to help these students working overtime to protect their planet.

In the meantime, let's do our math homework, shall we?

If you have...

Thirty minutes to learn the math: Read Bill McKibben's 2012 article "Global Warming's Terrifying New Math" in *Rolling Stone* (www.rollingstone.com/politics/news/global-warmings-terrifying-new-math-20120719).

Thirty minutes after that: Write an email to your high schooler's math teacher and/or principal and ask them to present this math problem to students. Or write a letter to the editor or a commentary in your local paper.

Ninety minutes: Watch the movie *Do the Math* by Kelly Nyks and Jared P. Scott or *DIVEST!* by famed *Gasland* director Josh Fox and Steve Liptay about the extraordinary Do the Math bus tour that launched the new divestment movement.

A classroom to teach: If you're a teacher, share this basic math with your students. Write a commentary. Editorial-page editors are always eager for new voices and innovative angles on hot topics. Educational journals often welcome new ways to bring climate information to students.

83. Divest. Get Everyone To.

Divestment is speeding up the clock on the final accounting that will show fossil fuels are out and clean energy is in.
— Lou Allstadt, former Mobil Oil senior executive

The woman in the flowing white gown made a lovely bride — unless, of course, you minded the crown of greasy black smokestacks on her head.

By holding a mock wedding, University of Oregon (UO) students were trying to convince the university to divest and dump their fossil-fuel stocks. Students had already held a nine-week-long sit-in at the president's office to no avail, so they decided to make the romance official. They issued invitations, assembled a wedding band, dressed as university officials, and marched their oily bride down the aisle.

Onlookers, invited to voice objections now or "forever hold your peace," prevailed on the groom — the faux UO president — to reconsider the union.

"Can't you see this relationship is toxic?" they shouted. The wedding made headlines. Five months later, UO announced its plans to divest.

Welcome to the biggest divestment movement in human history. Countless individuals and 850-plus institutions have divested more than $5.44 trillion.

"Just as we argued in the 1980s that those who conducted business with apartheid South Africa were aiding and abetting an immoral system," says Archbishop Desmond Tutu, "today we say nobody should profit from the rising temperatures, seas, and human suffering caused by the burning of fossil fuels."

And it's working. So well, in fact, that Energy Transfer Partners — owner of the notorious Dakota Access Pipeline — has sued Greenpeace and other groups for dampening investor enthusiasm for its dirty-energy projects. "The damage to our relationships with the capital markets has been substantial," complains the lawsuit. That's because it's not only colleges and churches dumping the dirty stuff. The Rockefeller brothers, the California Public Employees' Retirement System, the International

Monetary Fund, the British Medical Association, and even Ireland have piled on.

In other words, the world is starting to leave coal, oil, and gas at the altar.

It doesn't hurt that renewables are so cheap and that investors are realizing they stand to lose trillions in "stranded" fossil-fuel assets as the world turns from climate-killing dirty energy to clean. World Bank president Jim Yong Kim declared, "Sooner rather than later, [financial regulators] must address the systemic risk associated with carbon-intensive activities." He even told the *Guardian*, when calling for a carbon tax and the end of fossil-fuel subsidies, that he was impressed by the energetic divestment campaigns on US campuses. This is a stunning statement from the president of the world's largest development bank.

All of this has ExxonMobil shareholders nervous. That might explain why, at their 2017 meeting, they passed a resolution calling for disclosure of how public policy and consumer behavior will affect the assets of the world's largest oil company.

Let's keep them fidgeting by accelerating the flow of global capital away from dirty fuels and toward clean energy!

1. Flee your megabank. Member-owned, nonprofit credit unions are small, focused locally, and don't bankroll pipelines. Our cool credit union even donates bikes to students. To find an ethical home for your earnings, ask friends for recommendations and search these databases: BankLocal (http://banklocal.info) and SWICH (https://swich .to/businesses).

2. Divest from fossil fuels in your own investments. There's now even an S&P 500 Fossil Fuel Free Index (symbol SPYX). Invest in wind, solar, wave, or emissions-cutting technologies.

If you have . . .

Thirty minutes: Visit 350.org's divestment resource page (https://gofossil free.org/how-to-divest) or DivestInvest (http://divestinvest.org).

Read HIP Investor's "Resilient Portfolios & Fossil Free Pensions" (http://hipinvestor.com/wp-content/uploads/resilient-portfolios-and -fossil-free-pensions-byHIPinvestor-gofossilfree-vfinal-2014Jan21.pdf).

One hour: Open a credit union account. Help kids open one, and explain why it's important. Tell your megabank why you're breaking up with it.

Change to a fossil-free insurance company. As of this writing, only Allianz has dumped coal and upped investment in renewables.

Any time (all the time): Ask everyone you know — your friends, family, neighbors, and colleagues; your churches, workplace, bank, alma mater, investment company, and city, county, and state governments — to go fossil free. See my handy sample divestment letter in appendix A (page 257).

84. Dress Up Your Door

The job of the artist is to make revolution irresistible.
— Toni Cade Bambara

S omeone will hurl a rock through our window. Neighbors won't loan us
sugar anymore.

My insecurities battered me as I hung fabric leaves from the maple
trees by our sidewalk.

Inspired by both Occupy Wall Street and the approach of Thanksgiving, I set out to decorate our yard with a display that expressed gratitude for things we often take for granted. Sharpie in hand, I wrote on
flame-colored leaves the names of things my family appreciates that are
threatened.

The EPA. Social Security. Salmon.

High schoolers bustled by, eyes averted from the lady on the ladder.
A jogger surveyed my fake orange leaves and nodded politely. I soldiered
on until the names of 150 wonderful, endangered things swayed and spun
in the air above our sidewalk.

Affirmative action. Coral reefs. Public education. Habeas corpus.
The Clean Water Act. High-speed trains. Glaciers.

Neighbors, it turns out, found the project pretty cool. My kids weren't
sure. Once, arriving home, they lingered in the car until peers passed the
house with all that stuff in its trees. But no rock splintered our windows.
No hate letters arrived in our mailbox.

My children became less self-conscious. Spying from the living room,
they announced "customers" — those who slowed, grinned, and read the
leaves. My kids began to act quietly proud of our creative yard display.
I relaxed, and delighted in how the installation changed over time as the
living leaves fell away.

We went on to create another fifteen installations — each more interactive than the first — involving single-use bags, "pocket poems," food
for the homeless, prayer flags, and faux fracked-gas pipelines. All cele-
brated the Earth and the ways it renews the human spirit. All advocated
for a just, healthy, fun world. My kids, their cousins, and our neighbors
and friends all helped. Journalists sometimes interviewed us. People left

flowers and even donations with notes like, "Thanks for being the neighborhood's inspiring house!"

Try playing with public spaces you control. Once you start maximizing the visibility of your apartment or office door, yard, balcony, cubicle, bike, car, shopping bag, water bottle, or body, you might find it, as we have, exhilarating and remarkably freeing. Easy projects like the ones below can open the door to extraordinary possibilities.

If you have ...

Five minutes: Search "climate protest art" for an eyeful of possibilities. Or visit my website (www.marydemocker.com) for simple and affordable ideas.

Tie a ribbon on your wrist or arm as part of a campaign to stop a pipeline, clear-cut, or oil train, or to support fossil-fuel divestment, solar investment, or climate-recovery legislation. If the band's visible enough, people ask, offering an opportunity to share your thoughts and perhaps inspire others.

One hour: Make a sign for one of your public spaces or decorate a T-shirt — or your arm cast, wheelchair, or cane — with paint.

Make a voter registration station on your lawn. Print forms from the internet; include the local precinct address. Feeling flush? Stamp them.

Five hours: Collaborate with friends and family to adorn your space with positive, invitational, and celebratory suggestions. Live in a high-rise? Make your apartment door an artistic climate kiosk. Post petitions, rally invitations, photos of your patio garden, and inviting information on how neighbors can join you.

Attach an "End climate destruction!" banner to your bike, or paint your child's bike trailer with "Keep fossil fuels in the ground! ☺."

Tattoo "Heal the climate" on your forehead.

85. Empower Women and Girls Everywhere

The relevant question isn't population but, rather, what are people in power privileging over a habitable future?

— MEGHAN KALLMAN, cofounder of Conceivable Future

In December 1995, when I was eight months pregnant, the Northwest "timber wars" were raging, and I sat by myself at a crowded fund-raiser for forest activists, eager to learn how these monkey-wrenching tree-sitters planned to stop illegal old-growth clear-cuts.

The keynote speaker began by asking if we were vegetarian or "even better, vegan."

Hands (including mine) went up and he praised us. Then he shouted, "How many of you have no children?" At thirty-three weeks, I was unsure of the correct answer, but I decided to side with the Catholics on this one and lowered my hand.

"How many of you will *never* have children?" Raised hands surrounded me. Jubilant, he yelled, "*You people are my heroes!*"

The crowd cheered. I caressed my belly, wishing I could cover the tiny ears inside. As the speaker ranted — humans are Earth's cancer, first-world children are super-consumers who drive global warming, people should adopt, not breed — I felt like I was at a revival meeting with a scarlet *B* (for "breeder") flashing for all to see. *I* was responsible for Earth's destruction.

Nature called, so I paraded my pregnant belly up the aisle, then through a lobby filled with marines gathered for their Christmas ball in the adjacent auditorium. I entered the women's bathroom, where a woman in a ball gown nursed a newborn. Another mom! She noticed my belly, smiled, and asked when I was due and if this was my first. I asked if this was hers.

"She's my thirteenth," she beamed. Though stunned — *thirteen? seriously?* — I smiled and said, "You are very, very blessed."

I headed home disoriented by back-to-back encounters with two such extreme views on reproducing, yet convinced that somehow, in the large sweep of the universe, it's a sacred act to bring new life into the world — even a damaged world.

I've pondered this should-North-Americans-give-birth issue ever since and have settled on this: Examination of our first-world consumption is critical, but shame should be directed at leaders who enable industrial pollution, not at couples whose love leads — as nature clearly intended — to babies. Governments *can* build a low- or no-carbon-energy infrastructure, one that includes families within the sustainable systems we already know how to build.

Of course, it's complicated. Overpopulation is a real problem. So let's start by supporting women and girls everywhere who don't want (or can't afford) children, providing good health care, education, and free birth control. Literacy is the number-one factor in helping women control their reproductive lives and the best strategy to ensure the well-being, not only of families, but of community resilience within a developing economy. Girl power = Smart public policy.

Let's also offer free women's self-defense classes as well as sexual-consent education for everyone. And let's support couples who — despite cultural pressures to produce grandkids — hesitate to bring babies into an increasingly uncertain world.

Finally, let's support every policy that increases children's chances for every climate they inhabit — from the climate of the womb to the climate of the planet — to be safe and filled with love.

If you want to . . .

Explore the issues: Visit the site Conceivable Future (http://conceivable future.org), which brings "awareness to the threat climate change poses to reproductive justice."

Celebrate New Life: Watch the documentary *Babies* by Thomas Balmès.

Spend time with babies and toddlers, those spiritual teachers who remind us that life is sweeter when they're in it.

Support reproductive rights: Take action on behalf of reproductive-rights legislation. Loudly. Give money to Planned Parenthood, which honors women and children and their health.

Support organizations such as the Center for Reproductive Rights (www.reproductiverights.org), which helps women worldwide gain access to safe, affordable contraception and provides empowering conversations about sexuality and reproduction.

Support sex education: Enroll your kids in sex ed classes. (Art and I recommend the Our Whole Lives curriculum, www.uua.org/re/owl.)

Sponsor one girl's education in a developing nation. Visit Educating Girls Matters (www.educatinggirlsmatters.org), which collects websites from various organizations sponsoring girls' education worldwide.

86. Pivot Your Volunteering Skill Set

"Mothers Unleash Their Organizing Power on Climate."
— GRIST MAGAZINE HEADLINE AFTER 2016 PRESIDENTIAL ELECTION

For sixteen years, I've volunteered at my children's schools. I've served lasagna at fund-raisers, played school receptionist, helped families navigate enrollment paperwork, accompanied kid musicals on piano, made centerpieces for silent auctions, and joined committees, such as one formed to support a family in crisis. It's been an effective — and often fun — way to build relationships with teachers, school staff, and other families.

In 2012, I cofounded a 350.org chapter in Eugene. After a meeting one night, as I stacked chairs and washed dishes, it struck me that I'd done the exact same things for a school silent auction the previous month. Come to think of it, during that evening's 350 meeting, we'd made funding decisions, organized carpools to the capitol, and gathered permission slips for posting photos of schoolchildren on our website — all things I'd done as a parent volunteer in schools.

Climate volunteerism, I realized, is much like volunteerism for schools, clubs, or any other community group with high ideals and low bank accounts. We raise money through raffles, direct appeals, and T-shirt sales. We advocate for carbon recovery programs in front of the mayor the way parents advocate for antibullying programs in front of the principal. We chant "Go fossil-free!" at state capitols the way parents chant "Go team!" at lacrosse games.

Whether to support children's education or climate justice, we parent volunteers create Facebook events, invite friends, host speakers, send newsletters, announce important save-the-dates, then scheme about how to make those events successful. We set up chairs for speakers, then take them down later. We hang flyers, sew costumes, paint banners, organize carpools, staff booths, moderate discussions, and craft press releases. We listen, problem-solve, and experiment, building community around our values, and we grow to love people we didn't know before.

Planet-saving requires no special skills, especially for parents, who are already natural organizers and community-makers. If you've helped with any school, club, or church event — or even assisted your child's lemonade stand — you've got everything it takes.

Ten to fifteen minutes: Schools often publish "wish lists" of supplies. Ask a community or student climate group for theirs. Pick one item to donate (or add it to your shopping list) and post the rest on social media, encouraging friends to fill it.

One hour a month: Are your brownies best sellers at school bake sales? Donate some to the next climate fund-raiser — or give them to student leaders, college divestment groups, or burned-out leaders. Invite children to join in and decorate treats — or at least lick the bowl.

With kids, post flyers for an upcoming documentary, rally, or speaker panel. Designate roles — "Tack & Tape Holder" or "Keeper of the List" (the list of bulletin boards locations, checking them off as you go). When they were younger, my kids loved the challenge of rearranging crowded bulletin boards to fit our posters, whether for school plays or climate protests.

Two to five hours a month: Become a parent-volunteer coordinator for your local climate group. Welcome parents warmly and use a website such as SignUp.com (https://signup.com), as schools do, to easily plug busy parents into family-friendly events.

If you're media-savvy, administer a climate group's blog, Facebook page, or Twitter account. Ask your kids for advice on creatively publicizing upcoming events, as well as to help build the buzz. Then watch the fun.

Try buddying up with another parent; they cheer for your kid at the basketball game while you testify in city hall for the municipal oil train ban (or vice versa).

87. Level Up the Antibullying Campaigns

If twelve-year-old girls can be brave and stand up to bullies, we should expect the same from the adults who lead our country.
— LILAH, member of DC Bully Busters, age twelve

"Here comes Fish Lips!" the boys in the back of my bus announced gleefully. A slender, studious ninth-grader named Greg beelined through the hail of insults toward an empty seat. A hulky tenth-grader approached and hissed, "Get *out* of my seat."

Silently, Greg moved to a new seat. As I and the group of boys watched — and the bus driver didn't — the bully followed him to the new seat. "That's my seat, too. Get *out!*" he said, and socked Greg in the stomach. The moment is seared into me — the flat, early morning light, the diesel fumes, the boys heckling, Greg bent over, gasping, and my sudden nausea and terror.

Maybe this memory explains why I feared for my son on his first day of middle school. Forrest *was* greeted at the front door by a gang of older students, which would have been frightening if they hadn't been members of the "Where Everyone Belongs" club, which cheerfully greeted new sixth-graders.

This national movement recognizes bullying as preventable and harmful to both the target and perpetrator. It was embraced by our school, and my kids learned to be "upstanders, not bystanders."

Adults often play by very different rules, though, and children are often dismayed to find that antibullying programs end at the schoolyard fence. One group of eleven- and twelve-year-old girls was so appalled by modern politics and today's politicians that they started DC Bully Busters, a nonprofit organization dedicated to "teaching politicians what middle schoolers know about bullying."

We can apply the same principles to the powerful fossil-fuel bullies who terrorize the global playground. Bullying is characterized by a power imbalance, and there's none greater than between the mega-rich fossil-fuel industry and the rest of us. They've been taking our lunch money for decades — and it's time to stop them.

The fossil-fuel industry also has lackeys — a slew of institutes with names as misleading as the science they promulgate. These campaigns effectively stall critical conversations about our climate crisis. This allows fossil-fuel bullies to dangerously alter our atmosphere, feigning ignorance as to their culpability.

Through relentless misinformation campaigns, they confuse the public and intimidate and harass every scientist, journalist, and politician who dares challenge industry, whose only concern is profits. Unfortunately, many times, we find that the global playground monitors — our lawmakers — have been bribed to look the other way.

So let's take a cue from the DC Bully Busters.

"In fourth grade," one member said in a meeting with lawmakers, "we learned the Three *R*s of standing up to bullies. The first *R* is "recognize," where you realize that someone is a bully. The second *R* is "refuse," where you refuse to be a bystander. The third *R* is "report," and in school that might be telling a teacher. In politics that might be telling the voters or the media."

What do you say we apply the Three Rs to climate bullying?

If you want to . . . .

Overview bullying prevention: With your kids, visit DC Bully Busters (http://dcbullybusters.com). Try one of their ten ways to help end bullying in politics. Check out their teaching kits. Learn the Three *R*s. Ask kids how they link this to climate issues.

Know who the bullies are: Stay informed through up-to-the-minute research and exposés by organizations such as: Oil Change International (http://priceofoil.org), DeSmog (www.desmogblog.com), Union of Concerned Scientists (www.ucsusa.org), *Inside Climate News* (https://inside climatenews.org), Democracy Now! (www.democracynow.org), the *Guardian* (www.theguardian.com), and of course the wonderful wit of John Oliver (http://iamjohnoliver.com).

Discuss local climate politics and local politicians with kids. Are any of them climate bullies? How might you employ the Three *R*s to publicly call them out and try to stop their efforts?

88. Creatively Disrupt

I had a moral obligation to step up.
— JOHN CARLOS, referring to his now-iconic Black Power salute
  after receiving his 1968 Olympic medal

When my daughter graduated from high school, I took pictures and quietly wept. Valedictorians reminisced. Graduates performed poignant poems, songs, and dances. Administrators thanked parents, shared inside jokes, and encouraged graduates to "take the world by storm!"

It was all truly heartwarming. Except that no one mentioned what is, to me, the climate elephant in the auditorium. No one acknowledged that, unless we radically transform everything fast, the world will take our kids by storm.

Watching, I wondered, *What if graduation speeches powerfully addressed today's burning questions? What if solemn and joyous celebrations offered honest conversation, useful advice, and inspiration?*

A "what-if" angel heard me, apparently. Soon after, at my niece Katherine's UC Berkeley Law graduation, I witnessed valedictorian Tamila Gresham beautifully honor blacks fallen to police violence. She asked minority students to stand so everyone could cheer their achievement of graduating from a top law school despite society's race barriers. She respectfully challenged the largely white audience to affirm that Black Lives Matter. She was met with cheers, a standing ovation, and a sense of gratitude for speaking authentically.

Later, while officiating my niece's wedding, Tamila similarly used the occasion to link the love between Katherine and her now-husband Subin with their love of social justice. At both ceremonies, Tamila disrupted expectations in ways that honored both the moment and those gathered in celebration, guiding us to new visions of what's possible.

Creative disruption was common during the Question Authority era of my childhood, with its draft-card burnings, pie-in-the-face political protests, and world-rocking Black Power salutes on the Olympic medal stand.

Now, though, I parent in an Obey Authority era. Yet parents are great candidates for reviving creative disruption in ways that fit today's zeitgeist. Just as Tamila — as a minority student graduating from a prestigious law school — was uniquely positioned to discuss race, justice, and educational access, parents are uniquely positioned to discuss climate, justice, and our children's future. Families interact with countless institutions — schools, teams, after-school programs, medical offices, religious communities — from whom we can demand authentic conversations about our children's well-being.

In 2018, we're moving backward in that national climate conversation. Gag orders forbid some state employees from even saying "global warming" or "climate change," and President Trump barred EPA and Department of Agriculture scientists from speaking at all with the public or media. Moreover, the manufactured "debate" over the reality and causes of climate change has had the intended effect: Citizens censor themselves around a subject perceived as confusing and controversial.

The problem only gets worse the longer we dither. As Bill McKibben writes, "Winning slowly is exactly the same as losing."

Parents can lead here by reasserting our authority and refusing to be silenced. We can remove our gags at PTA meetings, football games, graduations, and weddings — as Tamila did — and lovingly insist on authentic climate conversations about protecting our kids.

Disruption takes courage. It risks alienating people, some of whom may claim they appreciate the message but not the delivery. This may always be the case, and each disruption should be well considered for its impact. But when the time, place, and delivery are right, disruptive messages can resonate powerfully, provide beacons of hope — and inspire others to speak out, too.

If you prefer to speak...

Silently, without speaking: Make family T-shirts saying, "We'd rather be on the bullet train!" Wear them on flights or road trips. Visit Brandalism, an international collective that replaces ads with artistic protests (http://brandalism.ch), and consider becoming a "brandal."

Print fossil-fuel warning stickers from Kathleen Dean Moore's website (http://www.riverwalking.com/special-projects.html). Some folks put them on gas pumps. You could plaster them on your own property, like your car!

Post fact sheets from Moms Clean Air Force around town that link fossil fuels and childhood diseases (www.momscleanairforce.org/armed -with-the-facts).

Out loud: See "Become a Victory Speaker" (page 187), "Shout from the Solar Roller Coaster" (page 192), and "Light Yourself on Fire" (page 248), among others, for ideas.

89. Support Grassroots Groups

Big impact change starts with the individual. No one else can bring what you have. You show up, you say yes, and then you bring your magic.
— Kelsey Cascadia Rose Juliana

In the wake of the 2016 presidential election, after Donald Trump and the GOP took over two branches of government, I received countless solicitations from environmental groups. You probably did, too. Each essentially said:

> Dear Mary,
>
> Wow, things were already bad! Now they're *so much worse*! Our [Big National Environmental Group] is doubling down on our work and needs your support more than ever! Please send us *as much money as you can*!
>
> Signed, [Big National Environmental Group]

I felt irritated with everyone trying to scare me into giving them money, and a bit sad that this was all they wanted. I craved an invitation to build a movement. This is the solicitation I wanted to receive:

> Dear Mary,
>
> Wow, things were already bad. Now they're so much worse. That's why our [Local Group] needs you more than ever. Please join us Saturday to share food and stories and figure out how we'll fix things together. We don't want your money — we want *you*. Bring your big ideas, passion, and if you can, food to share. And kids! We'll provide juice, paint, poster boards, and a couple of teenagers to help them unleash art while we organize the actions we'll take to resist tyranny and heal the climate.
>
> Gratefully, [Local Group]

Oh wait, I did receive something like that from 350 Eugene (though I think they *did* ask for money). As solicitations from national groups kept arriving, though, I learned that most Big Green environmental groups, despite promising names such as the World Wildlife Fund, Environmental Defense Fund, The Nature Conservancy, and Conservation International — to name some of the largest — don't actually advocate what top climate experts insist we need: to stop burning fossil fuels. Unbelievably, some even take money from or forge "partnerships" with Shell, ExxonMobil, Chevron, Monsanto, BP, and other polluters.

Don't believe me, though. Visit the websites of the above-mentioned Big Greens, or that of any group you encounter, and critique them using the following questions. Here's what I found with one of the biggest Big Greens, The Nature Conservancy: Do they advocate keeping global warming to below 1.5 degrees? (No.) Keeping coal, oil, and gas in the ground? (Nope.) Do they "partner" with polluters Dow Chemical and Shell Oil? (Why, yes!) Do they ask you to calculate *your* carbon footprint? (You bet!) Just for fun, challenge kids: Count the major polluters on a Big Green business council!

Parents must understand: We haven't won any major environmental victories since the Reagan era. The reason we keep losing ground, reported a 2012 study by the National Committee for Responsive Philanthropy (NCRP), is because wealthy donors mostly fund top-down, elite Big Green groups with DC offices. The Big Greens then distance themselves from the noisy, "confrontational" grassroots groups that demand true reform, despite the fact that according to the report, "a vocal, organized, sustained grassroots base is vital to achieving sustained change."

If green philanthropists don't understand this, the Koch brothers certainly do. The millions they've lavished on the Tea Party have helped build a powerful grassroots infrastructure that has effectively destroyed planet-saving policy for years. To reverse our long losing streak, NCRP recommends donors build a similar infrastructure, then give 20 percent of grant dollars to low-income groups and communities of color and 25 percent to "grassroots advocacy, organizing, and civic engagement led by the communities most affected by environmental ills."

So join or fund local grassroots environmental and social justice groups. Or find grassroots climate justice groups that demand policies supported by science: 100 percent clean and renewable energy by 2050, and the restoration of the atmosphere to the safe upper limit of — at most — 350 parts per million of carbon in the atmosphere. Make sure your chosen group has divested from banks that fund pipelines (most Big Greens actually have no policy against investing in fossil fuels themselves).

Let's invest our money and energy with effective grassroots groups. We need change more than ever, and as Gandhi said, "A body of determined spirits fired by an unquenchable faith in their mission can alter the course of history."

If you have...

Five minutes: End donations to groups that haven't divested from fossil fuels, that have polluting corporate representatives on their boards, or that are in a "collaboration" or "partnership" with polluters. To vet them, read their mission statements critically, and check out their "partners."

Money or time: Support or join national climate action groups that have no investments in coal, oil, and gas. These include: Climate Hawks Vote (http://climatehawksvote.com), Earth Guardians (www.earthguardians .org), Friends of the Earth (https://foe.org), Greenpeace (www.green peace.org), Honor the Earth (www.honorearth.org), League of Conservation Voters (www.lcv.org), the Hip Hop Caucus (http://hiphopcaucus .org), Rainforest Action Network (www.ran.org), Sierra Club (www .sierraclub.org), the US Climate Action Network (USCAN, www.us climatenetwork.org), and 350.org (https://350.org).

More money: Donate enough to help your group hire powerful youth/ women/minorities to build a vibrant, intersectional local climate justice movement right in your town.

90. Interrupt the Mars and Geoengineering Fantasies

Don't let anyone tell you we can escape to Mars; we couldn't even evacuate New Orleans.
— AL GORE

I filed into the science museum's IMAX movie with dozens of families carrying massive popcorn buckets and huge cups of soda. For the next forty-five minutes, Jennifer Lawrence's husky voice narrated what sounded like a "propo" film made by Lawrence's character Katniss Everdeen in *The Hunger Games*.

I thought the film, entitled *A Beautiful Planet*, would teach us how to protect the Earth. Instead, it implies that, since we can't stop burning fossil fuels, maybe we can...go to Mars! The film doesn't explain how we'd move 9 or 10 billion people to a dry rock 33 million miles away or how we'd breathe, drink, or grow food there, or that we'd never come home, since our emissions on take-off would render our "beautiful planet" unlivable.

For decades, the fossil-fuel industry has lied about the reality and causes of global warming, suppressing information about science-based solutions that call for humanity to stop burning fossil fuels. Now, even science museums are entertaining outlandish solutions that require — surprise! — burning *more* fossil fuels.

The Mars fantasy is a favorite of wealthy entrepreneur Elon Musk, who gets far more press than scientists like Mark McCaughrean, senior advisor for science and exploration at the European Space Agency. Regarding Musk's plans to colonize Mars, McCaughrean tweeted, "I'm less concerned about making humans a multiplanetary species than I am about making the Earth a sustainable multispecies planet."

Another fantasy is large-scale geoengineering, which intends to directly counter the effects of climate destruction. The most popular version involves lowering Earth's temperature by spewing particles like sulfur dioxide, alumina, or calcium carbonate into the stratosphere to block the sun's rays, an idea that's expensive, dumb, ugly, and unjust.

Expensive because it would be a colossal, perpetual job.

Dumb because it's foolish to mess with Earth's huge, complex natural systems. (We're already doing that now.)

Ugly because it dooms us to permanently hazy skies.

And unjust because climatologists predict it will further decrease rainfall in many regions, like Africa, already ridden with drought and famine.

Even Harvard scientists are pursuing large-scale geoengineering. According to *MIT Technology Review*, they hope to conduct an experiment in 2018 using the "StratoCruiser," which is funded in part by geoengineering sugar daddy Bill Gates.

How do world-class research institutions like Harvard not recognize the large-scale-tinkering-with-the-planet thing as obviously whacked? As Naomi Klein asks rhetorically about geoengineering in *This Changes Everything*, "The solution to pollution is…pollution?"

But there's another problem with these scenarios: The illusion of a fix, even an expensive and Stupid one, takes pressure off to behave more responsibly now. If there's a fix, why change?

As parents, we can debunk false solutions with our kids. Our problem is straightforward: too many greenhouse gases in our atmosphere. So is the solution: Cut emissions and draw those gases down to safe levels. Teach kids about all the ways we do that using: wind and solar power, carbon farming, smart glass, refrigerant management, methane digesters, regenerative grazing, alternative cements, solar highways, high-speed rail…to name just a few. Protecting the atmosphere we have is far easier than fundamentally altering it — or trying to manufacture a livable atmosphere millions of miles from home.

If you want to…

Know more: Read *Drawdown: The Most Comprehensive Plan Ever Proposed to Reverse Global Warming*, edited by Paul Hawken. This beautifully photographed, well-researched, accessible book is a catalog of Smart, doable climate solutions that governments and local communities can put in place now.

Be the voice of reason: Call Bill Gates at his foundation's DC office (202-662-8130) or tweet (@gatesfoundation) to demand he divest from geoengineering. Ditto for Harvard University. Call (617-495-1502), tweet

(@harvard), or email the president (president@harvard.edu) and demand its Solar Geoengineering Research Program not conduct tests in our atmosphere. Instead, they should redirect funds into safe, healthy, and just solutions.

Send letters challenging editors — especially at children's magazines and educational websites — when articles cheerlead geoengineering, adaptation (without mitigation), and the colonization of Mars.

91. Inoculate Everyone

Nobody likes to be misled, no matter their politics.
— Dr. John Cook

One day, several years ago, my husband opened a letter, then frowned. "Wow," he said, shaking his head, "this wacko is trying to recruit other wackos into supporting his crazy agenda."

The letter arrived from ultraconservative Art Robinson, who rails against evolution and government in his failed bids to represent Oregon in Congress. He calls public schools a "devastating form of child abuse." He has proposed sprinkling nuclear waste over our oceans. And he wanted my math- and physics-teaching husband's signature on the now-notorious Global Warming Petition Project, also known as the Oregon Petition, which claims there's "no convincing scientific evidence" for human-caused global warming. Apparently, Robinson had sent it to thousands of science teachers, regardless of their backgrounds in climate science.

My husband tossed the invitation, muttering, "He might actually get a few people who are easily duped." Indeed, the petition boasts 31,487 signatures, and it has often been trumpeted in the media and by industry-funded front groups as proof that, hey, the science can't be settled on global warming if that many *scientists* say so, right?

Except that the signatures aren't verified. Signatories have included Charles Darwin and characters from the TV show *M*A*S*H*. And only thirty-nine signatories even claim to be climatologists. *That's an eighth of a percent.* Most signatories may indeed have science degrees, but like my husband, they have no experience modeling long-term atmospheric changes.

The Oregon Petition is a worthless document. Yet, as intended, it confuses the public about the single greatest crisis facing our children. Worse, it helps confuse teachers, many of whom struggle with climate literacy themselves (see "Fight for Climate Literacy in Schools," page 142).

However, there's a silver lining to Robinson's well-funded lie: When people find out about this stunt, they're less susceptible to climate BS when they next step in it. In just two easy steps, we can use this to "inoculate" people against fake news:

1. Tell people that some politically motivated groups, like those who created the Oregon Petition, use misleading tactics to try to convince the public that there is a lot of disagreement among scientists about human-caused global warming. They hope to dupe the public about the need for urgent climate action. However, how credible is a document riddled with fake signatures — that, for example, one of the Spice Girls "signed" twice?

2. Offer this fact: Virtually all climate scientists — more than 97 percent — agree that humans are dangerously heating the planet.

Just those two steps, according to a University of Cambridge study, make people less likely to believe false climate information in the future.

According to a study by University of Queensland's John Cook, who has researched the effects of climate misinformation, students are intrigued by this dramatic story and want to hear details. And like a booster shot, the more often people hear it, the better it works. Given the scale of both the climate crisis and climate denial, it may, in fact, be the single-most-important inoculation parents can give.

If you have...

Five minutes: Inoculate your family: Show your old-enough kids the Global Warming Petition Project (http://www.petitionproject.org). Explain its aim to deliberately confuse voters about the climate crisis, which helps industry-backed climate deniers get into Congress. Describe the signature scandal and how this reveals that the petition's purpose is to mislead, not inform. For a good synopsis of the whole story, read DeSmog's "Oregon Petition" article (www.desmogblog.com/oregon-petition).

Thirty to sixty minutes: Inoculate educators and decision makers: Write an email outlining the two-step inoculation and send it to your child's science teacher, school's principal, or any representatives who deny the dangers of climate change. Inoculate readers of your local paper through a letter to the editor; post about the issue on social media.

Seven weeks, two to four hours weekly: With your teen, take a free online course, "Making Sense of Climate Science Denial" by Dr. John Cook through the University of Queensland, Australia (https://www.edx.org /course/making-sense-climate-science-denial-uqx-denial101x-5).

92. Battle to Win

Power concedes nothing without a demand. It never did and it never will.
— FREDERICK DOUGLASS

It's 3 AM and I'm staring at my toddler's rear end. After waking to the dreaded "Mommy, my butt itches," I shine the flashlight and spot a white thread diving between her little buns.

Pinworms. What a brilliant design. Children swallow their eggs, which hatch inside kids' intestines. Worms crawl out nightly to lay millions of eggs, which irritate sensitive skin, so kids scratch, getting eggs under their nails, which they stick in their mouths, and round it all goes again.

Eggs even float on air, so when parents sweep, mop, and launder, we're at risk of inhaling them (totally unfair). If we take the prescription worms-be-gone drug, we can't tell for a month if we've really killed them, and we imagine our bums are itchy, assume we're inhaling eggs, then wash and mop obsessively.

I did, anyway.

Why am I telling you this? Because I can't stop thinking about how similar pinworm infestations are to fossil-fuel infestations in democracy. We find the sneaky critters in our museums, regulatory commissions, Big Green groups, retirement portfolios, global climate negotiations, and, yes, running our government.

Parasites never leave without a fight. One year, we exchanged pinworms with another family seven times. Crazed, we finally all agreed, Jonestown-like, to take the nasty drug together and wage a blitzkrieg of showering, garlic-eating, butt-cream-slathering, floor-mopping, hand-washing, nail-clipping, and sheet-washing. It worked because we learned pinworm habits and vulnerabilities, then fought intelligently and relentlessly.

Which is exactly what we must do with our fossil-fuel infestors. Here's how, in six steps:

1. Observe parasite agendas: Pinworms want our bodies for breeding grounds. Fossil-fuel infestors — Shell, BP, ExxonMobil, Koch Industries, and the politicians they fund in Congress — want our public lands to pump, drill, and frack for profit.

2. Understand their hostility to "the commons": Pinworms aren't community-minded. Similarly, the right wing and its oily bosses attack "the commons" — the atmosphere, waterways, soil, and resources we collectively inherit and pass to our kids. It includes wonderful collective creations like libraries, parks, scientific research, social programs, public universities, and government itself.

    Infestors attack government when it protects the commons and try to eliminate Social Security, Medicare, voting, bargaining rights, public schools, public lands, public housing, and other hallmarks of democratic society.

3. Expect parasites to remain parasites: I've met many pinworms, but never a reformed one. You can't beg, partner with, or meditate away pinworms. ExxonMobil has had forty years — since learning that its product dangerously heats the planet — to embrace and accept the truth about climate destruction. It's still refusing.

4. Unite around a clear agenda: Working together is key to eliminating infestations, but action must be timely and goals clear. When we fight together — and strategically — we win. But fight we must.

5. Use the proper remedies: Lice shampoo won't cure pinworms. Complaining about infestors over beer or mocking them on late-night TV doesn't lessen their grip on our democracy. But organizing to flush them out for good *does*.

6. Practice lifelong preventative maintenance: Politics is just personal hygiene on a larger scale. It never ends. There are always parasites seeking hosts, whether in our bodies or our democracies. Defend both. When our bodies feed earthworms, we'll rest peacefully, knowing we've left our worldly treasures for our descendants to enjoy and protect.

If you have...

Ten to fifteen minutes: Visit OpenSecrets.org (http://www.opensecrets .org) to discover who funded your representatives in the last election. Tell them that accepting dirty-energy money loses your vote.

Visit the Center for Media and Democracy's PR Watch (www .prwatch.org) and learn about the Bradley Foundation, with triple the Koch brothers' spending power and a fraction of their visibility.

A few free evenings: Watch *Citizen Koch* by Carl Deal and Tia Lessen, and read *Dark Money* by Jane Mayer.

93. Break Human Laws
(to Obey Nature's Laws)

I came here to let them know that what they're doing is wrong.
— Water protector IYUSKIN AMERICAN HORSE of the Sicangu Lakota

Our teens lowered their textbooks. "Really? Jail? At Christmas?" Art and I traded glances. "Maybe."

I downplayed the parents-in-lockup scenario. "Look, we'll only risk arrest if the State Department recommends the president approve the Keystone pipeline."

Silence. I tried cheerfulness. "That might not happen!" They frowned.

I sighed. "You realize we have no other morally responsible choice?" They nodded slowly and returned to homework.

That was 2013. Big Oil was force-feeding us the Keystone XL pipeline, which NASA's Dr. James Hansen called "the fuse to the biggest carbon bomb on the planet." So we joined ninety-six thousand Americans who committed to risking arrest to end the madness. Climate destruction, we figured, would damage our kids more than one yuletide with Mom and Dad behind bars.

Jail wasn't our "home for the holidays" — President Obama rejected the pipeline's permit in 2015. But in 2017, the Trump administration resurrected the monstrosity, while also gutting environmental protections and inviting polluters to drill, frack, export, and burn to their hearts' content. Which means we will need to continue breaking human laws until those laws are changed.

At least we're in good company. Jesus taught his disciples to "turn the other cheek," which was a powerful form of nonviolent civil disobedience. Jesus asked others to demand change — and even breach laws — but without harming their oppressors.

Today, nonviolent civil disobedience has a long tradition, having been used by Thoreau, Harriet Tubman, Gandhi, and Martin Luther King Jr. as well as suffragists, union workers, and countless activists opposing wars, injustice, and ecocide. And it works. Nonviolent civil disobedience has played a role in countless successful social-change movements.

Sometimes, those successes cost individuals dearly. Nonviolent protesters and civil rights activists have been imprisoned, abused, vilified, and even killed. That's part of the power of nonviolent civil disobedience; it challenges society's moral complacency.

With planetary stakes now so high, many parents and individuals are turning to direct action. These include Tim DeChristopher, who disrupted an oil and gas lease auction in 2008; Dr. James Hansen, who locked himself to the White House fence in 2013; Sandra Steingraber, who blocked a fracked gas project in 2015; my friend Sandra Clark, who blocked oil trains at an export terminal in 2013; and Iyuskin American Horse, who locked himself to a digger at the Dakota Access Pipeline construction site in 2016. There are countless others.

Sometimes these protests successfully stop projects. Sometimes people suffer negative consequences. Of the thousands of climate activists arrested in recent years, many have served prison time or paid fines, while others had charges dismissed or performed community service.

The costs of resistance are currently rising. In 2017, twenty states proposed or passed laws suppressing the right to protest, prompting a United Nations' condemnation of this legal trend as "alarming and undemocratic."

In 2014, one out of six Americans reported being willing to risk arrest to stop new fossil-fuel infrastructure. Art and I are two of them. We believe in the merits and necessity, at times, of well-considered and strategic use of nonviolent civil disobedience. It represents the moral high ground in an unjust situation. This is what Tim DeChristopher meant when, as he was sentenced to two years in prison, he said: "At this point of unimaginable threats on the horizon...this is what love looks like, and it will only grow."

If you want to...

Know your rights and get training: Visit the Civil Liberties Defense Center (https://cldc.org), which offers "Know Your Rights" pamphlets, trainings, and legal support to front-line activists. Also, 350.org (https://350.org) offers training in nonviolent civil disobedience, including for those in support roles (who don't want to risk the paddy wagon).

Support threatened activists: Visit the Climate Disobedience Center, co-founded by Tim DeChristopher (www.climatedisobedience.org). Learn about the "climate necessity defense"; send encouragement or legal funds. Attend climate defendants' trials.

94. Make Polluters Pay

Cleaning up after yourself is a mark of civilization.
— BILL MCKIBBEN

One day, curious to know whether entrepreneurship paid better than window-washing, Zannie, nine, and Forrest, six, announced, "We're trying a lemonade business!" They bought supplies, calculated how much they could fill each cup and still profit (four ounces), sold out, cleaned up, and noted the time: three hours of labor, four-dollar profit. Divided by two, that was less than a dollar an hour.

They learned that they should either go back to window-washing or serve cheaper juice. It didn't occur to them to save time by leaving their garbage outside or increase profits by stealing juice to sell.

It was sad, then, over ensuing years, to introduce them to scaled-up lemonade stands in the form of corporate politics. They were dismayed to see companies spew their pollution and run, not understanding why adults would even *want* to pollute. But when they learned that coal, oil, and gas companies use the atmosphere as a carbon dump, then lie about global warming and get away with it — all with taxpayer subsidies — they sputtered, "That is *so* stupid!"

Stupid, indeed. As Bill McKibben writes, "The perverse logic of capitalism [is] that it's extremely profitable to pollute." And no industry pollutes better — while foisting its costs onto everyone else — than dirty-energy companies. In fact, only *one hundred companies* are responsible for 71 percent of human-caused emissions. The top villains? ExxonMobil, Shell, BP, and Chevron. It's hard to quantify costs like species extinction or the threat to life itself, but one International Monetary Fund study that tallied environmental and health costs alongside oil subsidies, concluded that taxpayers worldwide paid roughly 5.3 trillion dollars to help oil companies fill our atmosphere with greenhouse gases in 2015.

Thank heavens, then, for the movement to make polluters pay the real costs of their toxic product. When carbon is taxed, demand for dirty energy falls along with emissions. In the United States, we — outrageously — have neither a carbon tax nor regulation, but public demand for both keeps growing. Of the various approaches, the fairest is a revenue-neutral fee that invests the money in a clean-energy economy, trains displaced fossil-fuel workers to join it, and assists communities coping with climate impacts.

How can parents help? Build public support for carbon pricing. Tell everyone this: Well-designed carbon pricing *works*. British Columbia's tax slashed carbon emissions by 16 percent (in the same period, the rest of Canada's rose 3 percent). But pricing alone is no silver bullet; we also need to end oil subsidies, leave dirty fuels underground, and much more.

We also need to pay our debts. Ugandan activist Bernadette Kodili Chandia argues in the film *The Yes Men Are Revolting* that industrial countries owe a "climate debt" to poor countries who've hardly contributed to global warming but are the most battered by its effects. She insists countries like hers be repaid as eco-creditors and strengthened for the coming shocks; one study predicts that nine out of ten African farmers won't be able to grow food in coming decades. The Green Climate Fund, an international fund to "support the efforts of developing countries to respond to the challenge of climate change," aims to do just that.

But we need more. Let's demand that polluters pay for damages they've done to the climate and our kids' futures. Once they're held accountable for harms they've unleashed, polluters may find it more profitable to wash windows. Or invest in clean energy.

If you want to . . .

Identify top polluters: Visit Carbon Majors (http://carbonmajors.org), which traced the majority of global emissions to just ninety entities.

Visit Price Carbon Now's "Carbon Pricing 101" tutorial (http://pricecarbonnow.org/tutorial/carbon-pricing-101). Watch "The Story of Cap and Trade," which clarifies carbon pricing (http://storyofstuff.org/movies/story-of-cap-and-trade).

Learn about "public nuisance" lawsuits cities are using to hold polluters responsible for costly adaptations (www.scientificamerican.com/article/cities-sue-big-oil-for-damages-from-rising-seas).

Make polluters pay: Visit Our Climate (www.ourclimate.us), which engages youth around carbon pricing, or join Citizen's Climate Lobby campaigns (https://citizensclimatelobby.org).

Support state attorneys general investigating ExxonMobil's climate denial. Visit Climate Liability News (www.climateliabilitynews.org/2017/11/13/maura-healey-massachusetts-exxon-climate-investigation).

95. Link Eco-Justice to All Justice

They tried to bury us. But what they didn't know is that we're seeds and we're growing.
— RUEBEN GEORGE, manager of the Tsleil-Waututh Nation Sacred Trust Initiative

In May 2014, the original route proposed for the Dakota Access Pipeline ran ten miles north of the largely white community of Bismarck, North Dakota. By September, the US Army Corps of Engineers had rejected that route, citing, among other reasons, risk to residents' water supplies. Instead, the eleven-hundred-mile-long pipeline carrying Bakken crude oil was rerouted right through the watershed of the Standing Rock Sioux Tribe.

It's critical to understand the many different ways that extracting, transporting, refining, and burning fossil fuels disproportionately harms people of color and the poor. One involves the trampling of indigenous rights and treaties.

Another involves the placement of coal plants, oil refineries, and chemical plants in or near communities populated by minorities. According to the NAACP, "Race — even more than class — is the number one indicator for the placement of toxic facilities in this country." Industrial neighbors create toxic air, which disproportionately hurts children; Latino and African American children are three times more likely than non-Hispanic whites to die from asthma.

Poor communities are also more vulnerable to hurricanes and severe storms, which are occurring more often now. The infrastructure in low-income neighborhoods is more vulnerable in general, and these neighborhoods tend to be lower-lying and more prone to flooding. Moreover, in the aftermath of storms, the poor struggle more to rebound. They often lack insurance coverage, and critical social services and public transportation are often suspended.

In the developing world, millions of people suffer when heat waves kill crops and unleash hunger across large regions. In tropical countries, cholera and other infectious diseases can spread rapidly after hurricanes, earthquakes, or landslides knock out clean water supplies. Those who are most vulnerable often succumb to these secondary disasters.

All of this is why climate change has been called "slow violence." And those with lower incomes are especially vulnerable. The World Bank reports that disasters like floods and droughts are expected to push at least a hundred million people into poverty within the next fifteen years. *One hundred million people.* That's about a third of the US population. I hope this helps makes clear how fighting for eco-justice is also fighting for social and economic justice for all. Many people have taken guidance and inspiration from the Standing Rock water protectors. When Sioux tribal members camped next to the pipeline route to peacefully protect their water, other tribes joined. Eventually, the world noticed, and thousands of supporters joined the encampment, including movie stars, war veterans, and tribal delegations from Africa and South America. Tribal elders insisted that the campaign stay both prayerful and peaceful, despite the highly militarized and provocative police response to it. This important aspect of nonviolent civil disobedience (see page 234) lent powerful credibility to this indigenous-led resistance.

The first thing anyone desiring climate justice can do is to listen to those most impacted by environmental injustice and support them in the ways they ask for support. And if you and your family are part of an oppressed group struggling for equality, seek out the growing number of environmental justice organizations that may be eager and able to assist in your efforts (see resources below). To end systemic oppression, we need to recognize the different ways we are all affected, and we need a coordinated response.

Here are some suggestions for things we can do:

1. Recognize our privilege. As a rule, the wealthier we are, the bigger our carbon footprint tends to be, regardless of our politics.

2. Recognize that minority populations suffer from systematic oppression (that reflects the United States' legacy of slavery and colonialism). This oppression is often exacerbated by polluting industries.

3. Strive to understand intersectionality in the climate movement. Show up at the rallies, and support the causes, of other groups. Fighting for the rights of everyone is part of saving the planet. Marginalized communities already suffer more from climate chaos, so these struggles are inextricably intertwined.

4. Listen with humility. Learn how to be an ally, comrade, or accomplice (or whatever term you prefer).

5. Speak up to interrupt insensitive or abusive comments or behaviors. Interrupt others who marginalize certain groups. Often this is un-intentional, but it still needs interrupting, such as when privileged people — particularly older white men and oftentimes white activ-ists and organizers — talk over youths or people of color. Suggest, "Let's hear some other voices now."

6. Leave our comfort zones. Be willing to hear feedback about what we're doing or saying that's unsupportive, however unintentional it may be.

7. Get newsletters and information about the struggles of other groups than the ones we're primarily involved in. Learn about the work of others committed to equality and nonviolence.

8. Be inclusive with our actions and language. Call people what they wish to be called (if unsure, ask if they have a preference). Many news organizations and others no longer use the terms "illegal" or "alien" when referring to human beings. They instead use the more respectful and accurate terms "unauthorized" or "undocumented."

9. Be ready to learn. When unsure of an appropriate response, ask, "How can I help?"

We all deserve humanity and security. Research confirms that safety and equality within any community makes everyone happier, so working for climate justice isn't just an ethical thing to do. It's community build-ing, educational, and makes life richer for everyone.

If you want to...

Educate your kids: Befriend people outside of your usual group. Model for kids how to leave our comfort zones.

Read or listen to the news and stay current on the plights, campaigns, and victories of oppressed and marginalized communities. Then, in age-appropriate ways, introduce and discuss them with your kids over dinner. Make plans for being part of solutions.

Post some inclusive and empowering messages kids will see as reminders. Syracuse Cultural Workers (www.syracuseculturalworkers.com) sells artistic social-justice posters, lawn signs, and more. *Rethinking Schools* also has a great catalog of resources (https://www.rethinkingschools.org).

Support communities worldwide: Support groups like Environmental Law Alliance Worldwide (ELAW; www.elaw.org), which helps communities worldwide challenge environmental abuses and speak out for clean air, clean water, and a healthy planet.

Explore indigenous environmental perspectives: Check out Honor the Earth (www.honorearth.org) and any writings from its executive director, Anishinaabe economist Winona LaDuke; Indigenous Environmental Network (www.ienearth.org); Camp of the Sacred Stones (http://sacred stonecamp.org); and Last Real Indians (http://lastrealindians.com).

Join an environmental justice group: The Green Spotlight provides a valuable list of environmental justice groups, including multicultural green groups "by and for Latinos, Indigenous/Native Americans (and native Canadians), African Americans, and Asian Americans." (http://www .thegreenspotlight.com/2017/07/environmental-justice-groups).

Transform your community: If it doesn't already exist, push for a local task force in your city to reduce racial and economic inequality, particularly as it intersects with potential climate-action policies. Find and support local bilingual, multicultural agencies dedicated to the empowerment of marginalized communities. In Eugene, for example, the grassroots environmental justice group Beyond Toxics works with vulnerable communities "to ensure environmental protection and health for all."

Rock the vote: Register voters in swing states. Volunteer to help people navigate the increasingly confusing voter ID laws, and oppose these laws and any representative who supports them. Visit Spread the Vote (www .spreadthevote.org), National Voter Registration Day (https://national voterregistrationday.org), and the "next generation, multi-racial civil rights organization" Advancement Project (https://advancementproject .org/home/).

96. Run for Office

If you find politics dumb and boring, and politicians, too, then you just have to create your own party or platform.
— JÓN GNARR

Anyone feeling depressed by the state of our politics might be cheered up by the story of comedian and father Jón Gnarr, who became the unlikely mayor of Reykjavík, the largest city in Iceland. Gnarr started his campaign as a joke, intending only to mock politics-as-usual after Iceland's 2008 economic meltdown.

Calling his the "Best Party," he publicized a platform with inane promises such as free towels at public pools to highlight the hollow promises of career politicians. He stunned everyone — including himself — by quickly becoming a viable candidate, at which point he formed real policies grounded in being "nice." To the delight of voters and the sputtering outrage of the political establishment, he continued to clown around, upping his promises to include "a drug-free Parliament by 2020."

Though his campaign was wackily subversive, he pledged to take the job seriously if elected, which, ultimately, he was. During his four-year tenure, he angered conservatives by occasionally showing up to official occasions in drag, but, with his Best Party, he managed to push through policies to help the economy recover.

In his memoir, Gnarr calls for artists, musicians, actors, and anyone with imagination to take over the political theater. You don't have to be a comedian or performer, just passionate. For instance, in 2018, bookstore owner Becky Anderson decided to challenge incumbent Peter Roskam for his congressional seat in Illinois' Sixth District because she was appalled by the Republican's enthusiastic support of Trump administration policies.

In an email to supporters, Anderson wrote: "We've watched Donald Trump and Congressional Republicans fight amongst themselves about how to best rip health care away from millions of people. We've heard them proudly disrespecting women [and]...deny the reality of climate change, putting our children and grandchildren's futures at risk.

"As the co-owner of Anderson's Bookshops, a city councilwoman, and a mother, I can tell you — none of this is good for business, none of this is good for government, and none of this is good for families. I'm fed up."

Right on, sister! A lot of us are fed up. And it's great that, during this climate emergency, honest, ordinary people are going to change things. Parents who can't run for office can heartily support good people who do. Give them money, if you can, or give them time. Knock on doors, make calls, register voters. And thank them for trying to save the world.

If you want to consider politics yourself, it might help to remember one of Gnarr's motivations: He was unemployed. What started out as a gag protest became a day job that supported his family. Not a bad side benefit in our campaign to reclaim environmental democracy.

If you want to...

Inspire your kids: Have them read *Madam President: The Extraordinary, True (and Evolving) Story of Women in Politics* by Catherine Thimmesh. The great website A Mighty Girl (www.amightygirl.com) has more suggestions of political biographies for young readers.

As a family, design a candidacy over dinner. Ask kids: Party name? Campaign promises? Holidays? New laws? How would they shut down polluters? Make a two-minute campaign video.

Ask your child's teacher or principal to run this as a class civics exercise. How about a schoolwide mock election?

Inspire yourself: Read Jón Gnarr's short, laugh-out-loud memoir, *Gnarr!*. Or Bernie Sanders's *Our Revolution*, or *Unbought and Unbossed* by Shirley Chisholm, the first black congresswoman and presidential candidate.

Run for office: Throw your name in the ring for a local office. Check out Run for Something (www.runforsomething.net), which supports the under-thirty-five set's political aspirations. Or join a women's leadership incubator such as VoteRunLead (https://voterunlead.org), She Should Run (www.sheshouldrun.org), Emily's List (www.emilyslist.org), and Running Start (https://runningstartonline.org).

Or run for a nonpolitical position. Even ceremonial positions (like our town's annual Slug Queen) can provide a platform to raise awareness about important issues.

97. Sue the Grown-Ups!

If we win this case, it will be revolutionary for environmental law.
— Kelsey Cascadia Rose Juliana

"Kelsey Juliana and some other kids are suing the governor of Oregon and the president of the United States for destroying the climate."

In 2015, when I said this to a group of sixty rambunctious fifth-graders, they suddenly stopped fidgeting. Kelsey was from a nearby high school, and she was, and is, helping bring international attention to her generation's climate plight.

I'd been invited to the school to talk about the block-long faux pipeline my family and 350 Eugene had built to protest a proposed fracked-gas export pipeline through Oregon. As the fifth-graders set up chairs, the teacher confided, "We've been studying climate science all year, and they're a bit daunted. It'll be good for them to hear about positive actions they can take."

The students got a kick out of photos and descriptions of our huge, fake pipeline, but once I mentioned Kelsey, they wanted to know everything about her. Did she have a lawyer? How often did she go to court? And how did she even start?

Abandoning my planned talk, I dove into the whole story: It wasn't just Kelsey. Children from all fifty states, in fifty separate lawsuits, have sued their governors and state agencies. Another group has sued the president and federal agencies such as the EPA. Kelsey is suing both her governor *and* the president. Why? Because she doesn't think the adults in charge should be able to trash the climate children need to survive.

I said, "Kelsey and her friends are saying that government support for Keystone pipeline, fracking, and big subsidies for fossil-fuel corporations isn't just dumb because burning coal, oil, and gas destroys the climate; it's *illegal* because it violates children's constitutional rights."

The other teacher said, "We've been studying the Constitution all year!" She and her students were off, reviewing together what they'd previously learned about the three branches of government and how each branch is meant to keep each other in check so our democracy doesn't fall to tyranny.

"Remember," the teacher said, "when two branches of government — the president and Congress — violate the Constitution, the third branch, the courts, can stop them. Like when judges stopped schools from segregating students in the 1960s. Segregation violated black students' constitutional rights, and the Supreme Court — remember the *Brown v. Board of Education* case? — told the government to stop."

The kids were rapt, especially when we got to the great plot thickener — how Big Oil jumped into the fray as "intervenors," which meant that, as codefendants with the United States government, they could bring their war chest to fight the children.

I explained, "Kelsey wants to make the government do whatever it takes to keep the Earth healthy. Climate scientists say we must stop burning fossil fuels, which oil companies don't like because they'll make less money. That's why corporations like Shell, ExxonMobil, and Koch Industries are telling the judge that children don't have the right to a livable planet."

I described Our Children's Trust (OCT), the nonprofit organization supporting the lawsuits, which was begun by environmental lawyer (and mom) Julia Olson, who lives in Eugene. I invited the children to the courthouse for an upcoming hearing.

They wanted in. They asked their teachers if Kelsey could visit, and a few days later, she did and revved them up even more. She said, "I'm telling the judge everything I'll lose if the government doesn't protect the climate. Can you guys do the same thing outside the courthouse?"

By then, the kids would have sawed off a limb if she'd asked. With the teachers' help, they drew posters of things they loved in Oregon and stood to lose to global warming — favorite beaches, sledding in wintertime, and rivers where they fished for family salmon cookouts.

On the big day in April 2015, Mayor Kitty Piercy joined the students for their colorful "Children's Tribunal" on the courthouse steps. *PBS NewsHour* sent a second camera crew, and footage of the children holding up their hand-drawn posters opened the ten-minute national news feature about the groundbreaking case.

The following year, in April 2016, the same teachers brought their fifth-graders and took over the federal courthouse steps for another hearing. They were joined by a hundred high schoolers who marched in carrying a huge, inflated Earth and leading a call-and-response chant:

"What do we want? *Climate justice!* When do we want it? *Now!*" Their arrival at the courthouse set off triumphant cheers from the crowd.

This time, the federal government wanted Judge Thomas Coffin to dismiss *Juliana v. U.S.*, but Judge Coffin ruled in the children's favor. He wrote that endless climate debate "necessitates a need for the courts to evaluate the constitutional parameters of the action or inaction taken by the government." His language made clear that — hurray! — he agreed with the children on a key point: They deserve to have their day in court.

We may never know what compelled Judge Coffin to take that courageous, pioneering legal step, but we do know this: Families swarmed the courthouse that day. Twenty-one plaintiffs, ages eight to twenty-one, faced him in his courtroom. Hundreds of middle schoolers and high schoolers filled the courtrooms. Hundreds of elementary school kids rallied on the courthouse steps.

Even more young people showed up for the next hearing months later, when the government and oil companies appealed Coffin's decision. That time, Judge Ann Aiken affirmed Coffin's decision, ruling two days after Donald Trump's 2016 election that the children could indeed take the US government to trial for violating their constitutional rights.

"I have no doubt," Judge Aiken wrote, "that the right to a climate system capable of sustaining human life is fundamental to a free and ordered society.... Without 'a balanced and healthful ecology,' future generations 'stand to inherit nothing but parched earth incapable of sustaining life.'"

Wow! A US federal judge ruled, for the first time, that the atmosphere should be protected!

Before you pop the champagne, though, let me clarify: This doesn't mean the government must change its evil ways — not yet, anyway. But these rulings *do* mean that the kids get a chance — finally — to put the US government on trial for knowingly destroying the climate.

That could lead to the big prize: a court order forcing our government to make science-based plans for restoring carbon dioxide concentrations to the safe upper limit of 350 parts per million.

This brilliant strategy bypasses any climate-denying president and Congress completely. And it just might work. In fact, many environmental lawyers say it may be the only shot we have left on the legal front.

That's galvanizing families everywhere. At the state level, children have won favorable rulings in Washington, New Mexico, Oregon, and

Massachusetts, and they continue to work hard on cases pending in other states. At this writing, children in twelve countries, including France, India, Pakistan, and England, are pursuing ongoing cases.

Big Oil tiptoed out the back after Judge Aiken said they'd be going to trial. The Trump administration, meanwhile, is fighting the children tooth and nail — which is why we need hundreds of people, children and adults, on the courthouse steps every time there's a hearing. We want judges to hear our voices, too.

If you have...

Five to ten minutes: Watch the short videos I took in 2016 of the students arriving at the courthouse on my website (www.marydemocker.com); look under "Climate Action." Then, visit Our Children's Trust (www .ourchildrenstrust.org) and get on their mailing list, sign a petition supporting youth plaintiffs in various lawsuits, and learn about youth-led legal campaigns in twelve countries. Watch one of OCT's several short films about the youth, or read the widespread news coverage the cases have received, including spots on Bill Maher's show and Trevor Noah's *The Daily Show*, and articles in *National Geographic*, the *New York Times*, and *Rolling Stone*. Order a T-shirt. Most importantly, find a lawsuit near your family — and show up at public hearings, maybe even with colorful displays of support.

Twelve minutes: Watch Kelsey's TedX Salem talk (http://tedxsalem.us /kelsey-juliana). She describes the youth lawsuit and asks the audience to "applaud us — but join us."

Twenty-five minutes: Watch Bill Moyers's inspiring interview with legal scholar and parent Mary C. Wood, the legal genius who pioneered "atmospheric trust litigation" (www.youtube.com/watch?v=ktyq3vK6Ppg).

98. Light Yourself on Fire

Optimism is a political act. Those who benefit from the status quo are perfectly happy for us to think nothing is going to get any better. In fact, these days, cynicism is obedience.
— ALEX STEFFEN

A funny thing happens lately at parties when someone asks, "So, what do *you* do?" To my usual "I-teach-the-harp-and-parent-two-kids" reply, I've started adding, "and fight for a fossil-free future."

Whenever I reveal my secret identity — *Climate Mom, defender of planet Earth, crusader against fossil-fuel bullies and their slack-spine political minions* — New Deniers instantly surround me.

I don't mean those who deny humans are cooking the planet. No, New Deniers are a breed whose leitmotif is more chilling than "Drill, baby, drill." They deny that any action can help. They push back their chairs and hold forth on why resistance is futile.

Here's their litany, reduced: "Fossil fuel runs everything. Citizens United killed democracy. Everyone's too busy and kids are zombies. It's too late, we're frogs in a heating pot, our kids are screwed, and we'll never outspend Big Oil." New Deniers argue against self-exertion and explain why my efforts will fail.

At first, I tried convincing them otherwise, but it was like trying to rescue a drowning swimmer only to be submerged by flailing limbs. So I adopted a new approach: Listen and validate.

"Yes," I'd agree, "it's challenging and overwhelming — unprecedented, really — to confront the extinction of everything." This seemed to help people feel better, but they rarely took positive action or joined protests.

I felt stumped. How could I reach those trapped in catatonic despair?

In 2010, a Tunisian street vendor protested government oppression by setting himself on fire, and his heartbreaking act launched an international movement. For years, I've pondered how desperate, authentic acts become lightning rods for collective action.

What act might have such resonance? I even contemplated self-sacrifice, imagining the headline: "Frisbee Mom Desperate to Save Children's Future Self-Immolates."

Of course, I don't tolerate pain well, and I would never willingly traumatize my children. Obviously, I'd never do it. But one day, when depressed friends sang in the New Denier choir at a party, something in me exploded and I lit myself on fire — figuratively speaking.

In one long outpouring of optimism, I said, "We're alive at *the* most extraordinary point in human history. Climatologists say we can avoid catastrophe by slashing global emissions 8 percent yearly — *if* we start today. That means choking off new dirty-energy infrastructure — and Northwest activists are doing it! Kids aren't zombies, they're imitating *us*. Let's model how courage and love trump tyranny." I recounted historical examples, such as during World War II.

"We'll do it again!" I said. "Grassroots activism is at an all-time high! Divestment campaigns proliferate! Solar and wind prices keep plummeting! Whole countries are banning sales of gas-powered cars! The Pope is saying, 'Keep it in the ground'! I've got to try, so help me or at least get out of my way."

And you know what? One old friend, said, "Wow. Your passion's inspiring." Then laughed, and asked, "When's the next protest? I'm coming."

If you're talking about...

Fossil-fuel battles: Highlight victories from the Northwest's "Thin Green Line" (www.sightline.org/research/thin-green-line), which is stopping new dirty infrastructure and exports to Asia. Mention successful anti-fracking campaigns (www.americansagainstfracking.org).

Resisting the GOP agenda: Review and point out the successful activism spearheaded by Indivisible (www.indivisible.org), which includes squashing the repeal of Obamacare. Name the states, like California, that currently resist GOP-led federal policy changes.

Despair versus hope: Read the 2017 study "Believe You Can Stop Climate Change and You Will" (https://m.phys.org/news/2017-05-climate.html). People who believe they can make a difference take more pro-climate action than those without hope.

Remind everyone that our children, future generations, and all living things are counting on us. Go ahead, fight cynicism with passion. Ask others to fight as if everything — including our children's lives — depends on bold action. Because it does.

99. Okay, Okay, Shrink Your Carbon Footprint

Remember in the introduction, when I said the last thing we need is another book on personal eco-superheroism? Well, I stand by that. Definitely, for the next handful of years, we need every single voice demanding a fossil-free future from every level of government and all aspects of our society, including industry. It's key to prioritize system change.

But it's also true that lifestyle changes, especially in carbon-spewing North America, help cut emissions, train our children in more Earth-friendly ways of living, and just plain feel good. They certainly feel better than continuing our unsustainable lifestyles. So, go ahead, change that incandescent light bulb, and consider all the other everyday actions you might take from this list of possibilities.

Just don't choose between these and taking political action. Speak out first, and do these things if you've still got time and energy:

1. Buy efficient Energy Star appliances for your home. Smile at your lower utility bills.
2. Seal air leaks in your walls. Caulk around windows. Stay cozy.
3. Buy clean energy — wind, wave, solar — from your local utility company.
4. In winter, lower your home heater's thermostat. Install a timer to auto-lower at night. Wear sweaters; use blankets.
5. Ditto with your water heater.
6. If renting, research energy-incentive programs and ask your landlord to hop on board.
7. Get a solar water heating system. Water heaters use the most energy of family appliances.
8. Repair refrigerator leaks. Properly dispose of climate-killing fluids used as refrigerants.
9. Cover single-pane windows with storm windows. Or replace them.
10. In summer, open windows nightly to cool the house. Run the attic fan. Close windows daily.

11. Use heavy drapes and close them at night to help retain heat — or keep you cool.

12. Make sure door thresholds aren't drafty.

13. Close your fireplace flue.

14. Paint walls a light color to use less energy for lighting. Use eco-friendly paint.

15. Use low-flush toilets. Follow the mantra: "If it's yellow, let it mellow; if it's brown, flush it down."

16. Consider a composting toilet system: Turn your spoils to soils!

17. Install low-flow faucets and showerheads. Use faucet aerators. Fix drips.

18. Take short showers instead of baths.

19. Properly sort your glass, paper, and plastic for your local recycler.

20. Use low-phosphate cleaning supplies.

21. Keep grocery bags or boxes in your car or bike bag, along with a travel mug or two.

22. Air-dry dishes (skip the dishwasher dry cycle).

23. Hang laundry on a line or drying rack.

24. Insulate well, especially in your attic and walls.

25. Live in smaller spaces. Don't build or buy big. Don't always heat/cool little-used rooms.

26. Use motion sensors for outdoor lights — it reduces energy use and light pollution.

27. Don't build new private pools, especially heated ones. Join the Y or a city pool to swim.

28. Install a living roof; it insulates well, manages storm water, and starts conversations.

29. Don't water your lawn. Get drought-resistant plants.

30. Pave less; green living areas soak up storm-water runoff beautifully.

31. Keep bees. Or invest in a hive with friends. Collect honey as payment.

32. Wash clothes in cold water. Wash full loads. Don't overdry clothing. Keep the lint screen clean.

33. Install solar power.

34. Don't just clean out the fridge — dust behind it (dusty coils = more energy used).

35. Use cloth napkins. Use rags instead of paper towels. Bring back the linen handkerchief!

36. Buy or make cloth menstrual pads to save money, avoid chemicals, and go zero-waste.

37. Go paper-free. Or buy recycled products made from at least 30 percent postconsumer waste.

38. Scrape car windows instead of using the heater to melt ice. Idling = bad. Moving arms = good.

39. In summer, roll down windows for first two minutes to circulate the air. Don't idle with the A/C!

40. Be a Zen-like driver. Angry braking and accelerating eats gas. Remember: Hotheads heat the Earth.

41. Attach a trunk or trailer to your car, not a roof rack; wind drag decreases fuel efficiency.

42. Don't drive one day a week. Extend this Sabbath longer if you can.

43. Keep tires properly inflated (efficient tires = less gas).

44. Recycle motor oil. Nasty stuff, but it can be reused.

45. Travel by train, bus, carpool, bike, skates, feet.

46. Buy electric or hybrid cars — used, if you can.

47. Extend the life of your current car; maintain it well. Making new cars = big carbon spew.

48. Use apps to avoid traffic jams and all that idling.

49. Skip the tourist-in-space flight on Virgin's Galactic. Give the $250,000 you save to climate groups.

50. Fly economy, which fits more people per row than first class. Prince William does!

51. Avoid or reduce eating beef and/or dairy (cow farts and burps = unburned methane = very bad). Try a plant-based diet.

52. Buy local foods. Support farmer's markets. Grow your own, which avoids packaging.

53. Buy organic brews (commercial hops are liberally sprayed).

54. Drink tap or filtered water, not water bottled in unhealthy plastic.

55. Support organic farmers using permaculture methods (not carbon-intensive agri-businesses).

56. Buy shade-grown, fair-trade coffee and chocolate. Buy sustainably harvested wood, not tropical hardwoods that destroy rainforests.

57. Buy everything you can locally to support local business and decrease miles traveled by your food and gizmos.

58. Buy used or repurposed everything. Embrace simplicity — less to buy, store, insure, dust.

59. Buy quality, long-lasting clothes made from wool, linen, cashmere, silk, hemp, organic cotton, and bamboo.

60. Avoid clothes made of petroleum (nylon) or tropical wood pulp (rayon, viscose, lyocell).

61. Support ethical businesses that walk the talk and support environmental democracy.

62. Buy books locally. Skip Amazon. Better yet...

63. Borrow books from the library (even this one).

64. Request a green burial. No embalming involved. Just a biodegradable coffin or shroud.

65. Replace incandescent light bulbs with (cheap!) LEDs.

100. Consider Yourself Invited

Is this something that only I can do?
Is this the most important task I can take on at this time?
Is this a project that will bring me joy?
— KATHLEEN DEAN MOORE

One evening, during the reign of George W. Bush, I arrived at the house where my writing group met, plopped down on the floral couch, and opened my laptop. Another mom in the group sat next to me and asked how I was. Holly and I often groused about the state of things, politically speaking, so I answered, "Great, except that I'm depressed about everything Bush and his buddies are doing to destroy the planet. How 'bout you?"

Holly paused a moment and said, "I'm good. I'm finding that the more I engage in solutions, the less anxious about it all I feel." I don't remember what she was doing to engage, exactly, only her certainty that jumping in was the only thing that finally helped her feel better. I began to watch for ways to engage, too. It took another couple of years — I hadn't been very politically active since my student days in New York City, and I now had small children — but her words stayed with me.

I realized eventually that I wanted to write stories about my kids and the beauty and peril of the world, and I studied and practiced as I could while parenting. When I sent in my first unsolicited climate commentary, I had only one credential: I was a worried mother unwilling to watch stupid policies take down my kids' world. No one had asked my opinion, but I gave it anyway, figuring that if humanity is destroying itself, it clearly needs to hear from a lot more mothers. It was published, so I wrote another one. Through my writing, I began to meet others working in countless ways to build a fossil-free future and raise empowered kids, often in ways I'd never encountered. No one gave them permission to jump in. They just did because it made them feel better, and that inspired others to leap — or tiptoe — in, too.

I cordially invite you now. Please. Jump. In. Or tiptoe, it doesn't matter. Don't wait for phone calls or signs from God (though if you need them, consider the spate of natural disasters your divine summons). Just start. Elect yourself or follow someone you admire. Choose one thing. Today. I hope you'll find, as I did, that Holly was right. Doing *something* offers its own benefit. When I do whatever it is I feel called to do, I feel more energized, connected, and hopeful, especially when I work with other families or my own kids. Transforming global civilization is an exciting mission, but it's a pretty large one and impossible to do alone. Kids make it more meaningful and fun.

Just imagine if we could model for them what Kathleen Dean Moore suggests in the quote that opens this chapter. What if we tried to make a difference by doing what only we can do and — this is key — what brings us joy? Joyous efforts, ours combined with the millions of others worldwide, are building that bridge to a future in which children thrive, a future that's not just healthy and equitable, but likely to involve a good amount of laughter.

Thank you for coming on this journey with me and my family. Blessings to you and yours.

With love and gratitude,

Mary

Appendix A: Sample Divestment Letter

This is a sample letter you can use to ask others to divest from fossil-fuel-related companies. For more on this, see "Divest. Get Everyone To." (page 209). Personalize it in your own way and send to religious institutions, local governments, educational institutions, businesses, and your aunt Sarah.

Dear [Institution/Person you know]:

It's wrong to wreck the planet. It's also wrong to *profit* from wrecking the planet.

As a parent, I find fossil-fuel investments particularly immoral because the world's poor — disproportionately made up of women and children — are the least responsible for climate chaos, yet suffer first and worst from its effects.

Will you join the growing global movement to divest from corporations that recklessly — and knowingly — ruin our planet?

Shell Oil and ExxonMobil have known for decades that their product is dangerously heating the planet. Yet both, along with other fossil-fuel companies, funded misinformation campaigns about the reality and causes of global warming. It's long past time to revoke their social license.

Please, divest from the coal, oil, and gas companies incinerating our children's futures.

Fossil-free investment is good business: Fossil-free portfolios increasingly perform better than their fossil-fuel counterparts, which now grow more vulnerable as the carbon bubble expands.

By investing in solutions, you'll accelerate the desperately needed transition to a clean-energy future. With solar and wind prices plummeting worldwide, you'll also profit. Win for you, win for families!

For fossil-free investment information, please visit any of these sites:

Resilient Portfolios & Fossil-Free Pensions (http://hipinvestor .com/wp-content/uploads/resilient-portfolios-and-fossil -free-pensions-byHIPinvestor-gofossilfree-vfinal-2014 Jan21.pdf)

350.org, "How to Divest" (https://gofossilfree.org/how-to-divest)

DivestInvest (http://divestinvest.org)

Thanks for divesting today!
Sincerely,
[Your name here]

Appendix B: Sample Neighbor Letter

This is a sample letter for inviting neighbors to a disaster-readiness gathering. See "Host a Disaster Party" (page 161) for what this is and why it's important. Customize this letter as much as you need to reflect your local issues and your personal relationships.

> Hi neighbors!
>
> I hope this finds you well. Please come to [our home/name location] on [this date] for a ninety-minute gathering. We'll talk about how we might support one another in the event of a natural (or unnatural) disaster.
>
> Our neighborhood is vulnerable to [name the concern: oil train blast/flood/industrial accident/wildfire/mudslide]. I'm not even sure of all that we need to be prepared for, but we'll find out together!
>
> We're starting with [name the number of households you've invited]. We will follow a widely used template from Los Angeles on disaster preparedness called "5 Steps to Neighborhood Preparedness." If you'd like to preview this program, please visit http://5steps.la.
>
> The program will help our neighborhood to do the following:
>
> Organize our contact information.
>
> Identify special skills, like medical, firefighting, or ukulele-playing.
>
> Identify vulnerable community members, such as pets, babies, seniors, and so on.
>
> Create a simple plan in case of an emergency.
>
> You don't need to bring anything. We'll provide: handouts/pens/refreshments/childcare. If it's rainy, we'll meet at [name alternate location]; otherwise, just show up [repeat meeting details].
>
> Please let me know if you can come by [calling/emailing; provide contact info].
>
> Thanks, and have a safe weekend.
>
> Best,
>
> [Your name and address]

Appendix C: Writing a Press Release

If you have an event that you want to promote or publicize, write a press release. Give it to news organizations at least a week before your event and to partner organizations at least six weeks before your event (so they can include it on calendars). Press releases are usually sent as emails (so you're not asking people who don't know you to open attachments). It's worth the extra minutes to find the right reporters and editors whose interests or specialties fit your event, whether that's climate and the environment, local news, the arts, families, community, social justice, local events, and so on. Follow your press release up with a phone call to remind reporters about the event, and again the morning of the event. If it involves children, tell them; reporters are eager for youth perspectives — because audiences are eager for the same.

In the press release, link your event with a larger news event or a compelling social or political issue. This provides media with the "hook" that they'd use to present your event. In the example below, which I wrote at the start of Itzcuauhtli's silent strike, I highlighted the then-upcoming Paris climate summit, provided quotes from Itzcuauhtli, his brother, and mother, and highlighted actor Mark Ruffalo's comments (high-profile connections often get attention). I also emphasized the focus on climate science, and hyperlinked relevant information to make it easy for reporters to learn more. It's important to write a press release like a short article — one page long, if possible — to make the reporter's job easy. Often, they'll just copy your writing and print it nearly verbatim, so give them everything they need to get your story out into the world.

In general, press releases follow a fairly standard format and contain the following elements, though you can adjust these as necessary. Here is what you should consider:

Use letterhead or a logo, if you have one.
Start with the words "For Immediate Release" in bold and all caps.
Include the date.
Provide a contact person (or two), with their title, phone number, and email, along with the organization name and website (provide URLs or use hyperlinks).
Create a catchy headline.

Provide a subtitle that further summarizes or explains the focus.

In a single page, describe the event as succinctly as possible. Start by placing the city and state before the first sentence, and then focus on who, what, when, where, why, and how. Include quotes from key people and several "calls to action." Tell people how to help or get involved: attend an event, sign up, join an action, write to politicians, get more information, and so on.

At the bottom of the press release, three pound symbols (###) signal the end.

Below is an example of a press release I wrote for Itzcuauhtli Martinez and Climate Silence Now. You can use this as a template for your press release.

FOR IMMEDIATE RELEASE

Dec. 4, 2014

Contact: Tamara Roske, [phone number; email address]
Climate Silence Now, www.climatesilencenow.org

11-YEAR-OLD ON DAY 39 OF SILENT STRIKE
CALLING FOR CLIMATE ACTION

Urgent Plea for "Adults Who Love Their Kids to Stand Up for Our Future"
Is Resonating with Children Worldwide — and Actor Mark Ruffalo

Boulder, Colorado — Itzcuauhtli Martinez, an 11-year-old indigenous eco-rapper, is in week six of silence to demand science-based climate action. He asks why kids should "go to school and learn all this stuff if there is not going to be a world worth living in? The so-called 'leaders' are failing us. We now face a crisis that threatens everyone's future. I'm taking a vow of silence until world leaders take action. When I say world leaders, I mean us. Maybe it's up to youth." His site is receiving hundreds of thousands of hits, many from children in other nations. More than 1,000 will join his Dec. 10th Day of Silence.

Itzcuauhtli's spokesperson is 14-year-old brother Xiuhtezcatl, director of Earth Guardians (www.earthguardians.org) and a co-plaintiff in a youth climate lawsuit (www.ourchildrenstrust.org) the Supreme Court considers Dec. 5th. The brothers, raised in the Earth-honoring ceremonies of their father's Aztec culture, perform eco-rap worldwide (www.youtube .com/watch?v=78SLfsHGAm8) and have shared stages with Michael Franti, Nahko Bear, and Trevor Hall. After co-leading the 400,000-strong People's Climate March, Itzcuauhtli despaired that it may be too late to avoid runaway climate chaos. "I felt desperate. I had to do something drastic to change our future." Accusing leaders of being all talk and no action, he stopped speaking Oct. 27th. His mother, Tamara Roske, calls the response from children worldwide "huge and overwhelming," especially after Mark Ruffalo called the strike "brave and thoughtful." The actor wrote, "I am also made heartsick by your despair, little one. Your silence is a symbol of the silence that will come from doing nothing."

The sixth-grader says the recent US-China climate agreement "is not strong enough. Scientists say we must cap carbon in the next year. If we wait another 15 years, which is when China said they'd cap carbon, it's going to be too late." Itzcuauhtli calls on leaders to implement the planetary "prescription" written by top climate experts who outlined a recovery plan based on science, not politics — the same remedy demanded in Xiuhtezcatl's lawsuits against state and federal governments. Itzcuauhtli vows to continue his strike as long as he must and asks supporters to visit Climate Silence Now (www.climatesilencenow.org) to join his silence Dec. 10th "even for an hour! When the silence strike is complete, step into your role as leader. Speak up about climate change and never stop." He promises more actions leading up to the UN Climate Summit in Paris in Dec. of 2015, where leaders can make meaningful, binding agreements.

###

Appendix D: Kathleen Dean Moore's Letter to Grandparents

This letter by Kathleen Dean Moore appears in *Moral Ground: Ethical Action for a Planet in Peril*, which she edited with Michael P. Nelson. Use this as is by copying it — or cutting it out — to send to your own parents or in-laws who, of course, adore their grandchildren. Or use it as a template to create your own letter asking elders to join or support the climate justice fight *now*, while there's still time to make a meaningful difference in their grandchildren's lives. Many thanks to Kathleen for granting permission for its use here.

Okay. It's time for a frank talk with grandparents.

This is given: We love our children and grandchildren more than life itself. We would, in literal fact, do *anything* for them.

This is also given: Grandparents are in a powerful position to protect the children and grandchildren. The first asset we bring to the work is a set of skills, experiences, and knowledge gained over a lifetime of productive work. The second asset is political clout. We vote, and there are a lot of us. We donate to political campaigns, and we have a lot to give. The third asset is something that we have in abundance that no other demographic has enough of: we have time. Put these assets together, and grandparents command the power to shape the new world. We can make sure that our children and grandchildren inherit a planet that will sustain their health and nourish their freedom to make a good life.

How? By organizing our huge political power to elect officials who will get down to the most important business of protecting the life-sustaining systems of the planet. Knock on doors for these politicians, then hold them to account. If politicians dither or jabber or make excuses, send them home. Organize for truly fast and effective climate change legislation. Organize for solar power. Organize to keep poisons from the air, the water, the agricultural fields.

Start small. Madeline's Grandparents for Clean Electricity.

Fierce Grandmothers for Safe Food. Retired People for Redwing Swamps. The Council of Elders for Responsible City Government. The Grandparents' Coal Boycott.

Redirect already-existing resources. Tell AARP it's time to stop worrying about the health of our colons and golf games and start worrying instead about our legacy. Focus the church educational program on environmental toxins instead of the nature of heaven. Get the alumni association off the cruise ships and into the streets with banners.

Get started. At this stage in our lives, it mocks death to waste time. We work all our lives to provide for the future of our children and grandchildren. We cannot let it all slip away in our last decades. A life-sustaining planet, not an MP3 player or a plate of cookies, will be our last and greatest gift to the ones we love the most. Without that gift, all other gifts are meaningless, and the hugs of a grandparent become cynical jokes on the most beautiful little ones, who do not deserve the struggles they will face.

P.S. Let us hear no more talk about the extra entitlements of the elders — entitlements to year-round perfect weather, an annual trip to Las Vegas, low taxes, easy Sunday crosswords, reduced greens fees, and the world be damned. It may be true that because we have worked hard, we deserve to sit around and enjoy the fruits of our labor; but we owe those fruits also to a stable climate, temperate weather, abundant food, cheap fuel, and a sturdy government — all unearned advantages that our children will not have if we don't act. It's tragic and culturally dysfunctional if our lives culminate in a radical selfishness that makes us angry and bitter (because selfish desires can never be satisfied) rather than in the respect that we truly earn as people of wisdom and responsibility.

Appendix E: Preamble to the US Constitution

In chapter 52, "Teach Civics — Not Reading — to Kindergartners" (page 133), I suggest making a contest out of memorizing the fifty-two word Preamble to the Constitution. Not only is this surprisingly fun, but it's an energizing way for families to learn about one little-known, yet profound principle: Our constitution guarantees to "promote the general welfare" to "our posterity." *Posterity*, according to the Oxford dictionary, means "all future generations of people."

This is profound. Our Constitution states unequivocally that all future generations of people — our kids — have a right to both freedom and good lives! Now *that's* something worth sharing with families everywhere.

And now, the Preamble:

We the people of the United States, in order to form a more perfect union, establish justice, insure domestic tranquility, provide for the common defense, promote the general welfare, and secure the blessings of liberty to ourselves and our posterity, do ordain and establish this Constitution for the United States of America.

Acknowledgments

This book began as an idea I shared with the right people, who, like loving members of an extended family, helped nurture its full potential. I am deeply grateful for support from the following people:

Kathleen Dean Moore, for more than a decade of inspiration and mentorship, and for believing in both me and this book at critical junctures.

Bill McKibben, a guiding light for activists and writers worldwide, for generously agreeing to write the foreword.

Melissa Hart, for midwifing so many of my ideas into publication with her honesty, intelligence, and humor.

My agent, Jennifer Unter, for confidently guiding me through the shape-shifting publishing industry.

My editor, Jason Gardner at New World Library, for offering this book patience, respect, and a strong hand when it needed one.

Jeff Campbell, for editing with a precision and wisdom that allowed this book to become even better than the one I'd dreamed of writing.

Everyone at New World Library, for being the perfect fit, which I knew when I first walked in and saw staff members doing yoga in the break room.

Tracy Cunningham, for embracing my vision for the book's artwork.

My publicist, Monique Muhlenkamp, for enthusiastically joining the climate revolution.

My niece Hannah Peck, for unleashing her artistic superpowers on my sketches.

Naomi Kirtner and Jeff Goldenberg, for graciously letting me share their family's story in my chapter about grief.

Cam Fox, Jim Neu, Zach Mulholland, Abi Goldenburg, Wesley Georgiev, Steve Thoennes, and Eliot Schipper, for last-minute research assistance.

My writing buddy Sally Sheklow, for always knowing when to say, "Stop talking and write down what you just said."

My neighbor Jen Thoennes, for consistently brilliant feedback — sometimes given in pajamas — on nearly everything I've published over that last five years.

Catia Juliana, Tim Ingalsbee, Tivon Ingalsbee, and Kelsey Juliana, for unwavering friendship, gluten-free desserts, and helpful feedback on every idea I bring to the Juliana-Ingalsbee living room.

Bonnie Souza, Jim Neu, and Laurie Powell for reading the whole book over the holidays, flagging errors, and cheering me on.

Julia Olson, Meg Ward, Elizabeth Brown, and the legal warriors at Our Children's Trust, for inspiration — and teaching me so much about children's inalienable rights over pizza and beer.

Environmental law scholar Mary Christina Wood, for encouragement and a unique perspective that has profoundly influenced my thinking about climate justice and the role of family.

Earth Guardians founder Tamara Roske, for demonstrating how families can do intensive climate work with integrity and laughter. Her sons, teen climate warriors Itzcuauhtli and Xiuhtezcatl, for collaborating with me on "creative disruption" projects with joy and clarity.

Carrie Ann Naumoff and Susan Dwoskin, for sharing their unique curriculum for fifth-graders and applying it with such generosity and joy. Their principal, Tom Horn, for believing in both science and civics.

Editors Sy Safransky and Ann Wiens, for hiring an emerging writer.

Mesa Refuge, the Spring Creek Project, Lisa and Jay Namyet, Amy Minato, and Joy Archer for generously hosting writing residencies.

My 350 Eugene cofounders, Deb McGee and Patricia Hine, for their friendship, organizing genius, and gift of a cabin on their beautiful land as I wrote this book.

350 Eugene members and leaders, for four years of creative disruption, food made with love, and showing up — even in downpours — to manifest the world that's possible.

The University of Oregon Climate Justice League students, South Eugene High School's Earth Guardians/350 club, and 350 Eugene intern Phacelia Cramer, for the youthful energy taking the climate movement where it must go.

Valerie Brooks, Karin Sundenberg, Oregon Writer's Colony, Willamette Writers, Tsunami Books, Amy Minato, Charles Goodrich, Carly Lettero, and my many teachers and readers, for encouraging my voice.

Karen Rainsong, for creating my gorgeous website.

My harp students, for so much music over so many years.

Kate Doyle, Mimi Dvorson, and Jennie Sherlock, for talking through countless ideas on our many hikes.

Cindy Evans, Michelle Tompkins, Kaya Stern-Kaufman, Fred Niemeyer, Aaron Poor, Jennie Sherlock, Lori Maddox, Julie and Charlie Tilt, Alice Holmes, Bethany Drohmann, Tricia O'Neill, Rich Phaigh, Alicia Derby, Anne Tearse, Kim Hyland, Julie Fischer, Klaartje Broers, the Eugene Waldorf School, Sanborn Western Camps, and my parent group, for supporting my family through the years in countless vital ways.

Our block family — Jen, Steve, Emma, Ben, and Andrew Thoennes — for the open doors and hearts.

Mimi Dvorson and Nir, Alia, and Geffan Pearlson, for three decades of being there with love and laughter, no matter what.

The Peck and DeMocker clans, for being such good family.

My mother, Janice DeMocker, for keen editorial feedback and, along with my father, John DeMocker, giving me a childhood blessed with open water, music lessons, and a grand sense of freedom.

Countless climate justice leaders, authors, investigative journalists, scientists, teachers, attorneys, artists, musicians, spiritual leaders, and community organizers, for sharing the many good works that guide this book and give me strength for the journey.

My husband, Art, for cheerfully nurturing me, this book, and our children with clarity, intelligence, and steadfast love.

Alexandra and Forrest, for transforming me into their mother, and then letting this book take over dinner conversations and tell our stories.

Endnotes

Introduction

Page xxi, *"The technology of community":* Bill McKibben, "How Close to Catastrophe?" *New York Review of Books,* November 16, 2006, http://www.nybooks.com /articles/2006/11/16/how-close-to-catastrophe.

Page xxi, *in 2015, Pope Francis wrote an encyclical letter addressed to:* Pope Francis, *Laudato Si: On Care for Our Common Home,* encyclical letter (May 24, 2015), section 114, http://w2.vatican.va/content/dam/francesco/pdf/encyclicals /documents/papa-francesco_20150524_enciclica-laudato-si_en.pdf.

Page xxviii, *philosopher Kathleen Dean Moore writes that "although environmental":* Mary DeMocker, "If Your House Is On Fire: Kathleen Dean Moore on the Moral Urgency of Climate Change," *The Sun,* December 2012, https://www.thesun magazine.org/issues/444/if-your-house-is-on-fire.

1. Get Clear on Why There's Hope

Page 3, *"prescription" for balancing the climate by 2100: TRUST Oregon,* Our Children's Trust video, accessed October 11, 2017, http://www.ourchildrenstrust.org /short-films.

Page 3, *if we accomplish those three things, climate stability is achievable:* James Hansen et al., "Scientific Case for Avoiding Dangerous Climate Change to Protect Young People and Nature" (Ithaca, NY: Cornell University Library, 2011), https://arxiv.org/abs/1110.1365.

Page 4, *those same scientists say humanity will lose its opportunity:* Ibid.

2. Herd Your Family Together

Page 5, *three main culprits, the first being central heating:* Tom Geoghegan, "What Central Heating Has Done for Us," *BBC News Magazine,* last updated October 1, 2009, http://news.bbc.co.uk/2/hi/uk_news/magazine/8283796.stm.

Page 5, *Completing the trifecta is the electronic device:* Jim Taylor, "Is Technology Creating a Family Divide?" *Psychology Today,* March 13, 2013, https://www.psychology today.com/blog/the-power-prime/201303/is-technology-creating-family-divide.

3. Plant Trees!

Page 7, *"the most powerful solution available to address global warming":* Paul Hawken, ed., *Drawdown: The Most Comprehensive Plan Ever Proposed to Reverse Global Warming* (New York: Penguin Books, 2017), 111.

Page 7, *it sequesters a surprising amount of carbon:* Ibid., 134.

Page 7, *nine-year-old Felix Finkbeiner described their simple strategy:* Plant-for-the-Planet

website, accessed October 11, 2017, https://www.plant-for-the-planet.org/en
/about-us/aims-and-vision.

Page 8, *keeping carbon locked up in our greenery:* Hawken, *Drawdown,* 109–11, 114–17,
128–34.

Page 8, *the United States is the biggest importer of tropical hardwoods:* "Avoiding Unsustainable Rainforest Wood," Rainforest Relief, accessed September 24, 2017,
http://www.rainforestrelief.org/What_to_Avoid_and_Alternatives/Rainforest
_Wood.html.

Page 9, *Rainforest Alliance, a certification organization*: Sarah Shemkus, "Better Ba-
nanas: Chiquita Settles Lawsuit over Green Marketing, But the Legal Battle
Isn't Over," *The Guardian,* December 19, 2014, https://www.theguardian.com
/sustainable-business/2014/dec/19/chiquita-lawsuit-green-marketing-bananas
-water-pollution. See also Justin Rowlatt and Jane Deith, "The Bitter Story Behind the UK's National Drink," BBC News, September 8, 2015, http://www.bbc
.com/news/world-asia-india-34173532.

4. Lift Moods and Grades

Page 10, *the risk of death by stranger abduction is far lower than:* For more information on
these dangers, see Lenore Skenazy, "Crime Statistics," Free-Range Kids, accessed
November 28, 2017, http://www.freerangekids.com/crime-statistics; "10 Leading
Causes of Death by Age Group, United States – 2015," Centers for Disease Control and Prevention, National Center for Injury Prevention and Control, accessed
November 28, 2017, https://www.cdc.gov/injury/wisqars/pdf/leading
_causes_of_death_by_age_group_2015-a.pdf; and "2015 US Lightning Deaths,"
The Weather Channel, July 16, 2105, https://weather.com/storms/severe/news
/united-states-lightning-deaths-2015.

5. Cultivate a "Used-Is-Cool" Culture

Page 12, *whose real wages haven't risen since 1978:* Drew Desilver, "For Most Workers,
Real Wages Have Barely Budged for Decades," Pew Research Center, October 9,
2017, http://www.pewresearch.org/fact-tank/2014/10/09/for-most-workers
-real-wages-have-barely-budged-for-decades.

6. Give Up One Thing

Page 15, *higher carbon footprint than perennial tree fruits:* Hawken, *Drawdown,* 66–67.

Page 15, *The American Red Cross has been criticized:* David Crary, "For Red Cross, Hurricanes Bring Donations and Criticism," Associated Press, September 14, 2017,
http://www.detroitnews.com/story/news/nation/2017/09/14/hurricanes
-red-cross/105599640.

7. Ditch the Diaper

Page 16, *24.7 billion annually in the United States alone:* DiaperFreeBaby, accessed August 1, 2017, http://diaperfreebaby.org.

Page 16, *"Disposable diapers should be considered as one"*: Rosalind C. Anderson and Julius H. Anderson, "Acute Respiratory Effects of Diaper Emissions," *Archives of Environmental Health* 54, no. 5 (1999): 353–58, http://www.tandfonline.com/doi /abs/10.1080/00039899909602500.

Page 17, *Diapers introduce bulk:* Whitney G. Cole, Jesse M. Lingeman, and Karen E. Adolph, "Go Naked: Diapers Affect Infant Walking," *Developmental Science* 15, no. 6 (November 2012): 783–90, doi:10.1111/j.1467-7687.2012.01169.x.

Page 17, *that number had dropped to 4 percent:* Erica Goode, "Two Experts Do Battle Over Potty Training," *New York Times,* January 12, 1999, http://www.nytimes .com/1999/01/12/us/two-experts-do-battle-over-potty-training.html.

8. Give Palm Oil the Back of Your Hand

Page 19, *Crisco's culprit — also in processed foods:* Meghan Telpner, "If Vegetables Don't Make Oil, What's Crisco?" https://www.meghantelpner.com/blog/vegetables -dont-make-oil-so-whats-crisco-really-made-of.

Page 19, *what* Scientific American *calls "the other oil problem":* "Stop Burning Rain Forests for Palm Oil," *Scientific American,* December 1, 2012, accessed October 13, 2017, https://www.scientificamerican.com/article/stop-burning-rain-forests-for -palm-oil.

Page 19, *torch the lands underneath:* Ibid.

Page 19, *Critics accuse RSPO…the least-offensive production methods:* Annisa Rah- mawati, "Cleaning Up Deforestation from Palm Oil Needs More Than Green- wash," *The Guardian,* June 5, 2014, https://www.theguardian.com/sustainable -business/deforestation-palm-oil-more-greenwash-greenpeace.

Page 20, *Learn palm oil's many aliases:* Ashley Schaeffer Yildiz, "Palm Oil's Dirty Secret: The Many Ingredient Names for Palm Oil," Rainforest Action Network, September 22, 2011, https://www.ran.org/palm_oil_s_dirty_secret_the_many _ingredient_names_for_palm_oil.

9. Bag the Cow

Page 21, *A 2009 World Watch Institute study:* Robert Goodland and Jeff Anhang, "Live- stock and Climate Change," *World Watch,* December 2009, http://www.world watch.org/files/pdf/Livestock%20and%20Climate%20Change.pdf.

Page 22, *Taxpayers pay…$400 billion in expenses:* Christopher Hyner, "A Leading Cause of Everything: One Industry That Is Destroying Our Planet and Our Ability to Thrive on It," *Georgetown Environmental Law Review,* October 23, 2015, https://gelr.org/2015/10/23/a-leading-cause-of-everything-one-industry -that-is-destroying-our-planet-and-our-ability-to-thrive-on-it-georgetown -environmental-law-review.

Page 22, *Impose a carbon fee on animal agriculture:* Hawken, *Drawdown,* 40.

Page 22, *cow backpacks are used by some ranchers:* Alex Swerdloff, "Cattle Farmers Are Fighting Climate Change with Fart-Collecting Backpacks for Cows," Munchies,

August 24, 2016, https://munchies.vice.com/en_us/article/vvqz4b/cattle
-farmers-are-fighting-climate-change-with-fart-collecting-backpacks-for-cows.
See also Beth Balen, "Global Warming: Backpack Captures Cow Farts," Guardian
Liberty Voice, April 16, 2014, http://guardianlv.com/2014/04/global-warming
-backpack-captures-cow-farts.

Page 22, *feed additives to reduce farts and burps:* Charlotta Lomas, "Farmers Fight Cow
Farts to Protect the Climate," *Deutsche Welle*, March 27, 2013, http://www
.dw.com/en/farmers-fight-cow-farts-to-protect-the-climate/a-16702813.

Page 22, *The USDA…other forms of protein production:* "USDA Funds Insect Farming
Research and Insect Based Food by Company All Things Bugs," PR Web, July 13,
2015, http://www.prweb.com/releases/cricket/flour/prweb12840894.htm.

10. Turn Fido into a Climate Hero

Page 25, *"to supply their energy needs, they will break down":* Margaret Gates, "Answers:
What Exactly Is an 'Obligate Carnivore'?" Feline Nutrition Foundation, last updated November 18, 2017, http://feline-nutrition.org/answers/answers-what
-exactly-is-an-obligate-carnivore.

11. Cut Your Toilette in Half

Page 26, *When researchers publicized that all women feel badly:* "Women of All Sizes Feel
Badly about Their Bodies After Seeing Models," University of Missouri-
Columbia, ScienceDaily, March 27, 2007, https://www.sciencedaily.com/releases
/2007/03/070326152704.htm.

Page 26, *the ugly side of an unregulated multibillion-dollar industry:* "When 'Pure'
Is Not Pure or How Did Our Lack of Regulations Allow Cancer-Causing
Chemicals in Baby's Products," Environmental Working Group, March 12, 2009,
http://www.ewg.org/enviroblog/2009/03/when-pure-not-pure-or-how-did-our
-lack-regulations-allow-cancer-causing-chemicals.

Page 26, *which can contribute to depression, self-harm, eating disorders, and suicide:*
"Healthy Body Image: Tips for Guiding Girls," Mayo Clinic, accessed October 13,
2017, https://www.mayoclinic.org/healthy-lifestyle/tween-and-teen-health
/in-depth/healthy-body-image/art-20044668. See also "Teen Depression," Mayo
Clinic, accessed October 13, 2017, https://www.mayoclinic.org/diseases
-conditions/teen-depression/symptoms-causes/syc-20350985.

Page 27, *Research and share the benefits of body hair:* Janell M. Hickman, "The Surprising
Benefits of Having Pubic Hair," October 15, 2017, https://www.glamour.com
/story/benefits-of-having-pubic-hair.

Page 27, *higher incidences of sexually transmitted diseases:* François Desruelles, Solveig
Argeseanu Cunningham, and Dominique Dubois, "Pubic Hair Removal: A Risk
Factor for 'Minor' STI such as Molluscum Contagiosum?" *Sexually Transmitted
Infections*, March 19, 2013, http://sti.bmj.com/content/89/3/216.info.

12. Shower Bikers with Praise (Not Puddles)

Page 29, *Researchers have found that kindness makes:* Jeanie Lerche Davis, "The Science of Good Deeds," WebMD, accessed November 6, 2017, https://www.webmd.com /balance/features/science-good-deeds#5.

Page 30, *This reduces "doorings," a common and sometimes:* Steve Annear, "To Avoid 'Doorings,' Cyclist Wants Drivers to Do the 'Dutch Reach,'" *Boston Globe*, September 8, 2016, https://www.bostonglobe.com/metro/2016/09/08/this-cyclist -wants-drivers-dutch-reach/V2Ei5bEiOCfU6ubxX1r8VN/story.html.

14. Keep the Clunker – Until You Can Afford Cleaner Transportation

Page 35, *Cradle to grave, they're half of the environmental impact:* Rachael Nealer, David Reichmuth, and Don Anair, "Cleaner Cars from Cradle to Grave," Union of Concerned Scientists, November 2015, http://www.ucsusa.org/sites/default /files/attach/2015/11/Cleaner-Cars-from-Cradle-to-Grave-full-report.pdf.

Page 35, *A lot of used Nissan Leafs and other EVs are available:* Doug DeMuro, "Holy Crap, Used Nissan Leafs Are Incredibly Cheap," Jalopnik, November 19, 2015, https://jalopnik.com/holy-crap-used-nissan-leafs-are-incredibly-cheap -1743475298.

Page 35, *by 2016 had pushed auto debt to an all-time high of $1.2 trillion:* Gwynn Guilford, "American Car Buyers Are Borrowing like Never Before — and Missing Plenty of Payments, Too," *Quartz Media*, February 21, 2017, https://qz.com/913093 /car-loans-in-the-us-have-hit-record-levels-and-delinquencies-are-rising-fast-too.

15. Don't Scurry, Be Happy

Page 37, *The* New York Times *advises, "The biggest single thing":* Justin Gillis, "Short Answers to Hard Questions about Climate Change," *New York Times*, updated July 6, 2017, https://www.nytimes.com/interactive/2015/11/28/science/what -is-climate-change.html.

Page 37, *When researchers asked children what they most wish for:* "Job Stress Can Affect Your Children," *NBC News*, last updated January 25, 2007, http://www.nbcnews .com/id/16795195/ns/health-livescience/t/job-stress-can-affect-your-children.

16. Bring Paris to You

Page 39, *To visualize that, imagine standing on Manhattan's:* Bill Chameides, "Picturing a Ton of CO_2," Environmental Defense Fund, February 20, 2007, http://blogs .edf.org/climate411/2007/02/20/picturing-a-ton-of-co2.

Page 39, *The other 45 percent keeps orbiting and trapping heat:* "Effects of Changing the Carbon Cycle," NASA Earth Observatory, accessed October 10, 2017, https://www.earthobservatory.nasa.gov/Features/CarbonCycle/page5.php.

Page 39, *impact of airplanes is* two to four times *greater:* "Air Travel and Climate

Change," David Suzuki Foundation, accessed October 10, 2017, http://david suzuki.org/issues/climate-change/science/climate-change-basics/air-travel -and-climate-change.

Page 39, *Air travel is…projected to double globally:* "New IATA Passenger Forecast Reveals Fast-Growing Markets of the Future," International Air Transport Association, October 16, 2014, http://www.iata.org/pressroom/pr/pages/2014 -10-16-01.aspx.

Page 40, *carbon offsets — criticized for commodifying pollution:* Duncan Clark, "A Complete Guide to Carbon Offsetting," *The Guardian,* September 16, 2011, https://www.theguardian.com/environment/2011/sep/16/carbon-offset -projects-carbon-emissions.

18. Get Electrified

Page 44, *traps* eighty-six times *more heat in the atmosphere than carbon:* Mark Hand, "Court Tells Trump's EPA to Enforce Methane Rule for Oil and Gas Drillers," *Think Progress,* August 1, 2017, https://thinkprogress.org/appeals-court-rules -against-epa-on-methane-rule-delay-8329e79d9d8e.

Page 44, *you just can't prevent climate-killing leaks when you drill for:* Tony Phillips, "US Methane 'Hot Spot' Bigger than Expected," *Science Beta,* NASA, October 9, 2014, https://science.nasa.gov/science-news/science-at-nasa/2014/09oct_methane hotspot.

Page 44, *particularly from fracking, which poisons water:* Abrahm Lustgarten, "Are Fracking Wastewater Wells Poisoning the Ground beneath Our Feet?" *Scientific American,* June 21, 2012, https://www.scientificamerican.com/article/are-fracking -wastewater-wells-poisoning-ground-beneath-our-feeth.

Page 44, *"worse than coal and worse than oil.":* A. R. Brandt et al., "Methane Leaks from North American Natural Gas Systems," *Science* 343, no. 6172 (February 14, 2014): 733–35, http://science.sciencemag.org/content/343/6172/733.

Page 44, *Induction stoves, for example, are 73 percent more efficient:* "Induction Cooking — 73% More Efficient than Gas," Electrolux, June 8, 2010, http://newsroom .electrolux.com/us/2010/06/08/induction-cooking-73-more-efficient-than-gas.

Page 45, *We're also eyeing heat pumps, which use half the electricity:* "Heat & Cool: Heat Pump Systems," US Department of Energy, accessed October 14, 2017, https://energy.gov/energysaver/heat-pump-systems.

Page 45, *The number-one indicator that someone will buy solar panels:* Kevin Dennehy, "Why Solar Adoption Can Be Contagious," Yale School of Forestry & Environmental Studies, December 3, 2014, http://environment.yale.edu/news/article /why-solar-adoption-can-be-contagious.

Page 45, *push representatives to adopt, low- or no-emission vehicle programs:* James Ayre, "US Department of Transportation Reveals $55 Million 'Low or No Emissions' Bus Grant Program," CleanTechnica, September 18, 2017, https://cleantechnica .com/2017/09/18/us-department-transportation-reveals-55-million-low-no -emissions-bus-grant-program.

19. Sleep Tight

Page 49, *I read…who discuss the concept of "life energy"*: Vicki Robin, Joe Dominguez, and Monique Tilford, *Your Money or Your Life: 9 Steps to Transforming Your Relationship with Money and Achieving Financial Independence* (New York: Penguin Books, 1992), 55.

24. Be the 3.5 Percent

Page 61, *two hundred environmentalists, mostly indigenous, were murdered worldwide:* "Defenders of the Earth: Global Killings of Land and Environmental Defenders in 2016," Global Witness, accessed October 10, 2017. https://www.globalwitness .org/en/campaigns/environmental-activists/defenders-earth.

Page 61, *or even "pure scum":* Marianne Lavelle, "Steve Bannon's Trip from Climate Conspiracy Theorist to Trump's White House," *InsideClimate News*, November 16, 2016, https://insideclimatenews.org/news/16112016/steve-bannon-trump -white-house-climate-conspiracy.

Page 61, *Honduran environmental activist Berta Cáceres:* Elisabeth Malkin and Alberto Arce, "Berta Cáceres, Indigenous Activist, Is Killed in Honduras," *New York Times*, March 3, 2016, https://www.nytimes.com/2016/03/04/world/americas /berta-caceres-indigenous-activist-is-killed-in-honduras.html.

Page 61, *experienced violence and abuse from police and private security firms:* Julia Carrie Wong, "Dakota Access Pipeline: 300 Protesters Injured After Police Use Water Cannons," *The Guardian*, November 21, 2016, https://www.theguardian .com/us-news/2016/nov/21/dakota-access-pipeline-water-cannon-police -standing-rock-protest.

Page 62, *people who believe they've seen a ghost:* Michael Lipka,"18% of Americans Say They've Seen a Ghost," Pew Research Center, October 30, 2015, http://www.pew research.org/fact-tank/2015/10/30/18-of-americans-say-theyve-seen-a-ghost.

Page 62, *And in this period of US resistance politics:* Natalie Broda, "Civil Unrest, Protest Numbers, at All Time High Says OU Poli-Sci Prof," *Oakland Press News*, February 7, 2017, http://www.theoaklandpress.com/general-news/20170207/civil -unrest-protest-numbers-at-all-time-high-says-ou-poli-sci-prof.

26. Learn the Fossil Fuel Nitty-Gritties

Page 66, *Fossil-Fuel Recipes: Coal:* "Coal Basics," Energy Kids, US Energy Information Administration, accessed October 30, 2017, https://www.eia.gov/kids/energy .cfm?page=coal_home-basics.

Page 66, *Fossil-Fuel Recipes: Oil & Natural Gas:* Anne Rockwell and Paul Meisel, *What's So Bad about Gasoline?* (New York: HarperCollins, 2009), 8–10.

Page 67, *First, people dug up just a little…Then we dug up:* "History of Energy Consumption in the United States, 1775–2009," US Energy Information Administration, accessed October 16, 2017, https://www.eia.gov/todayinenergy/detail .php?id=10.

Page 67, *Coal pollutes the most, making smog:* "There Is No Such Thing as 'Clean Coal,'" Natural Resources Defense Council, March 2008, accessed October 16, 2017, https://www.nrdc.org/sites/default/files/coalmining.pdf.

Page 68, *Those greenhouse gases trap heat:* "What Is the Greenhouse Effect?" NASA Climate Kids, accessed October 16, 2017, https://climatekids.nasa.gov/review/greenhouse-effect.

Page 68, *"Why is the U.S. exporting fossil fuels to China?":* Timothy Gardner and Nina Chestney, "U.S. Coal Exports Soar, in Boost to Trump Energy Agenda, Data Shows," *Reuters*, July 27, 2017, https://www.reuters.com/article/us-usa-coal-exports/u-s-coal-exports-soar-in-boost-to-trump-energy-agenda-data-shows-idUSKBN1ADoDU.

Page 69, *Drilling for it leaks methane, which traps 86 times more heat:* Gayathri Vaidyanathan, "Leaky Methane Makes Natural Gas Bad for Global Warming," *Scientific American*, June 26, 2014, https://www.scientificamerican.com/article/leaky-methane-makes-natural-gas-bad-for-global-warming.

Page 69, *And fracking poisons our water & causes earthquakes:* "Hydraulic Fracturing 101," Earthworks, accessed October 16, 2017, https://www.earthworksaction.org/issues/detail/hydraulic_fracturing_101. See also Calvin Sloan, "Frackers Knew: Fossil Fuel Industry Has Known since 1967 That Injection Wells Cause Earthquakes, Despite Denials," Center for Media and Democracy's PR Watch, November 23, 2016, https://www.prwatch.org/news/2016/11/13178/frackers-knew-fossil-fuel-industry-has-known-1967-injection-wells-cause.

27. Look to the Sky

Page 70, *This is partly because the solar-energy industry...driving demand up and prices down:* Many of these solar-power details are from Hawken, *Drawdown*, 9–10.

Page 70, *Germany makes more than twice as much solar power:* "Germany Is Burning Too Much Coal," *Bloomberg*, November 14, 2017, https://www.bloomberg.com/amp/view/articles/2017-11-14/germany-is-burning-too-much-coal.

Page 70, *Morocco built the world's largest concentrated solar installation:* Phoebe Park, "World's Largest Concentrated Solar Plant Switches On in the Sahara," *CNN*, February 8, 2016, http://www.cnn.com/2016/02/08/africa/ouarzazate-morocco-solar-plant/index.html.

Page 70, *China installs a soccer field–size set of panels every hour:* Beth Gardiner, "Three Reasons to Believe in China's Renewable Energy Boom," *National Geographic*, May 12, 2017, https://news.nationalgeographic.com/2017/05/china-renewables-energy-climate-change-pollution-environment.

Page 70, *and even built one solar farm in the shape of a panda:* Trevor Nace, "China Just Built a 250-Acre Solar Farm Shaped like a Giant Panda," *Forbes*, July 25, 2017, https://www.forbes.com/sites/trevornace/2017/07/25/china-just-built-250-acre-solar-farm-shaped-giant-panda/#42568c564685.

Page 70, *Solar energy costs are now as cheap as — and increasingly cheaper:* Katherine Bleich and Rafael Dantas Guimaraes, "Renewable Infrastructure Investment

Handbook: A Guide for Institutional Investors," World Economic Forum, December 2016, 4, http://www3.weforum.org/docs/WEF_Renewable _Infrastructure_Investment_Handbook.pdf.

Page 70, *Forward-thinking utilities are implementing "grid flexibility":* Hawken, *Drawdown*, 30–31.

Page 70, *Molten salt tanks, for example, are game-changers:* Ibid., 15.

Page 70, *in 2021 Los Angeles plans to unveil an eighteen-thousand-battery:* Ibid., 33.

Page 71, *Third-party leasing...on our economy and democracy:* Ibid., 11–12.

Page 71, *Cradle to grave, they're still far cleaner than any fossil fuel:* David Biello, "Dark Side of Solar Cells Brightens," *Scientific American*, February 21, 2008, https://www.scientificamerican.com/article/solar-cells-prove-cleaner-way-to -produce-power.

Page 71, *all renewable energy — is in the Grand Oil Party's crosshairs:* Dino Grandoni, "The Energy 202: Pruitt Plays to GOP Base by Repealing the Clean Power Plan," *Washington Post*, October 10, 2017, https://www.washingtonpost.com/news /powerpost/paloma/the-energy-202/2017/10/10/the-energy-202-trump-plays -to-gop-base-by-repealing-the-clean-power-plan/59db9d5530fb0468cea81e16.

Page 71, *Public utility commissions sound boring:* Joby Warrick, "Utilities Wage Campaign against Rooftop Solar," *Washington Post*, March 7, 2015, https://www .washingtonpost.com/national/health-science/utilities-sensing-threat-put -squeeze-on-booming-solar-roof-industry/2015/03/07/2d916f88-c1c9-11e4 -ad5c-3b8ce89f1b89_story.html.

28. Be a WIMBY (Wind in My Backyard!)

Page 72, *William threw himself into building a windmill:* William Kamkwamba and Bryan Mealer, *The Boy Who Harnessed the Wind: Young Reader's Edition* (London: Puffin Books, 2016).

Page 72, *Plus, wind power just... "demand from coast to coast":* Hawken, *Drawdown*, 3, 4.

Page 73, *It's the fastest-growing industry period:* Bureau of Labor Statistics, "Fastest Growing Occupations," *Occupational Outlook Handbook*, accessed October 13, 2017, https://www.bls.gov/ooh/fastest-growing.htm.

Page 73, *reduced noise and bird deaths, a common concern:* Hawken, *Drawdown*, 3.

Page 73, *The greatest support for new wind farms occurs when: local communities:* European Commission, "Public Support for Wind Farms Increases with Community Participation," *Science for Environment Policy* 387 (September 25, 2014), http://ec.europa .eu/environment/integration/research/newsalert/pdf/windfarms_support _community_participation_Sweden_energy_387na1.pdf.

29. Be Here Now

Page 75, *"Future generations...have no voice":* William Aiken et al., "A Buddhist Declaration on Climate Change," Forum on Religion and Ecology at Yale, May 14, 2015, http://fore.yale.edu/files/Buddhist_Climate_Change_Statement _5-14-15.pdf.

30. See the Human Face of Climate Impacts

Page 76, *That extra .5 degree signified a death warrant:* Jessica Leber, "How the Islands That Will Be Destroyed by Climate Change Won a Victory in Paris," Fast Company, December 21, 2015, https://www.fastcompany.com/3054791/how-the -islands-that-will-be-destroyed-by-climate-change-won-a-victory-in-paris.

Page 76, *"The poor — both those living in poverty and those just barely":* "Climate Change Complicates Efforts to End Poverty," World Bank, February 6, 2015, http://www.worldbank.org/en/news/feature/2015/02/06/climate-change -complicates-efforts-end-poverty.

Page 76, *Rather than keep waving the tired image of an iceberg-stranded polar bear:* Kate Manzo, "Beyond Polar Bears? Re-envisioning Climate Change," *Meteorological Applications* 17, no. 2 (June 2010): 196–208, doi:10.1002/met.193, http://online library.wiley.com/doi/10.1002/met.193/full.

Page 77, *raised money to send delegates from the Marshall Islands:* "Victory!: The End of Fossil Fuels Has Begun," Avaaz, accessed October 11, 2017, https://secure.avaaz .org/en/climate_story_loc/?pv=352&rc=fb.

Page 77, *He announced a High Ambition Coalition of:* Karl Mathiesen and Fiona Harvey, "Climate Coalition Breaks Cover in Paris to Push for Binding and Ambitious Deal," *The Guardian*, December 8, 2015, https://www.theguardian.com /environment/2015/dec/08/coalition-paris-push-for-binding-ambitious -climate-change-deal.

Page 77, *Other Marshallese described seeing the remains:* Ari Shapiro, "For the Marshall Islands, the Climate Goal Is '1.5 to Stay Alive,'" *NPR*, December 9, 2015, https://www.npr.org/sections/parallels/2015/12/09/459053208/for-the-marshall -islands-the-climate-goal-is-1-5-to-stay-alive.

Page 77, *they pledged to keep global temperature increases "well below":* Chris Mooney and Joby Warrick, "How Tiny Islands Drove Huge Ambitions at the Paris Climate Talks," *The Washington Post*, December 12, 2015, https://www.washingtonpost .com/news/energy-environment/wp/2015/12/11/how-tiny-islands-drove-huge -ambition-at-the-paris-climate-talks.

31. Get the Facts

Page 79, *but that science was scrubbed in 2017 under EPA administrator:* Lisa Friedman, "E.P.A. Scrubs a Climate Website of 'Climate Change,'" *New York Times*, October 20, 2017, https://www.nytimes.com/2017/10/20/climate/epa-climate -change.html.

Page 79, *encyclopedia "anyone can edit" is the seventh most-visited website in the world:* Nicole Torres, "Why Do So Few Women Edit Wikipedia?" *Harvard Business Review*, June 2, 2016, https://hbr.org/2016/06/why-do-so-few-women-edit-wikipedia.

Page 79, *Yet Wikipedia doesn't have enough editors to effectively vet:* Will Oremus, "Wikipedia's 'Sockpuppet' Problem," *Slate*, October 23, 2013, http://www .slate.com/blogs/future_tense/2013/10/23/wikipedia_sockpuppet _investigation_is_paid_editing_the_problem_or_the_answer.html.

Page 79, *including Wikipedia's special vulnerability to "sock puppets"*: "Wikipedia Blocks Hundreds of 'Scam' Sock Puppet Accounts," *BBC News*, September 2, 2015, http://www.bbc.com/news/technology-34127466.

Page 79, *I recently checked every citation under Wikipedia's entry for "natural gas"*: On November 5, 2017, I clicked on and analyzed each of 119 citations on Wikipedia's "Natural gas" entry, https://en.wikipedia.org/wiki/Natural_gas.

Page 80, *editors are 91 percent males under twenty-six*: Laura Beck, "Wikipedia's Editors Are 91 Percent Male Because Citations Are Stored in the Ball Sack," *Jezebel*, January 25, 2013, https://jezebel.com/5978883/wikipedias-editors-are-87 -percent-male-because-citations-are-stored-in-the-penis.

Page 80, *by documenting California's twenty-one thousand known leaks*: Ingrid Lobet, "Two Ways Natural Gas May Be Escaping at Your Meter," inewsource.org, November 3, 2017, https://inewsource.org/2017/11/03/natural-gas-leaking.

Page 80, *by including the growing global resistance to fracking*: India Bourke, "Preston New Road: How a Fracking Protest Became a Movement," *New Statesman*, July 29, 2017, https://www.newstatesman.com/politics/energy/2017/07/preston -new-road-how-fracking-protest-became-movement.

33. Change Your Media Diet

Page 85, *When we watch shows...aren't as likely to partake in community activities*: Robert Kubey and Mihaly Csikszentmihalyi, January 1, 2004, "Television Addiction Is No Mere Metaphor," *Scientific American*, https://www.academia.edu/5065840 /Television_Addiction_is_no_mere_metaphor.

Page 85, *Some shows can even imperil our health...than people who skip the news or watch positive or neutral news*: Graham C. L. Davey, "The Psychological Effects of TV News," *Psychology Today*, June 19, 2012, https://www.psychologytoday.com /blog/why-we-worry/201206/the-psychological-effects-tv-news. See also Chaya S. Piotrkowski and Stephen J. Brannen, "Exposure, Threat Appraisal, and Lost Confidence as Predictors of PTSD Symptoms Following September 11, 2001," *American Journal of Orthopsychiatry* 72, no. 4 (Oct. 2002): 476–85, http://dx.doi.org/10.1037/0002-9432.72.4.476.

Page 86, *Americans consume an average of ten-plus hours of media daily*: John Koblin, "How Much Do We Love TV? Let Us Count the Ways," *New York Times*, June 30, 2016, https://www.nytimes.com/2016/07/01/business/media/nielsen -survey-media-viewing.html.

34. Build a Wall – of Inspiration

Page 87, *Abby Brockway, who blocked explosive oil trains*: Associated Press, "Oil Train Protesters: 5 Arrested for Blocking Tracks at Everett Rail Yard," *The Oregonian*, September 2, 2014, http://www.oregonlive.com/pacific-northwest-news/index .ssf/2014/09/oil_train_protesters_5_arreste.html.

Page 87, *Sandra Steingraber, jailed for fifteen days*: Website of Sandra Steingraber, accessed October 10, 2017, http://steingraber.com.

36. Try Protest Therapy

Page 91, *Famed psychologist James Hillman… "get ourselves in order":* Sy Safransky, "Conversations with a Remarkable Man," *The Sun*, July 2012, https://www.the sunmagazine.org/issues/439/conversations-with-a-remarkable-man.

Page 92, *the American Legislative Exchange Council literally writes:* "What Is ALEC?" Center for Media and Democracy, last updated October 13, 2017, https://www .alecexposed.org/wiki/What_is_ALEC%3F.

38. Dig, Dawdle, and Dog

Page 97, *Dogs are a natural antidepressant, as are all pets:* Kara Mayer Robinson, "How Pets Help Manage Depression," WebMD, accessed January 1, 2018, https://www .webmd.com/depression/features/pets-depression#1.

Page 98, *people even pay for laughing classes:* "What Is Laughter Yoga and How Can It Help You?" Laughter Yoga University, accessed October 13, 2017, https://laughter yoga.org/laughter-yoga.

Page 98, *A third of Americans are sleep-deprived:* Pam Fischer, "Wake Up Call!" Governors Highway Safety Association, accessed August 8, 2016, http://www.ghsa.org /resources/wake-call-understanding-drowsy-driving-and-what-states-can-do, 5.

Page 98, *people with spiritual communities report feeling happier:* Sally Quinn, "Religion Is a Sure Route to True Happiness," *Washington Post*, January 24, 2014, https://www.washingtonpost.com/national/religion/religion-is-a-sure-route -to-true-happiness/2014/01/23/f6522120-8452-11e3-bbe5-6a2a3141e3a9_story .html?utm_term=.979bc46bafob.

Page 99, *Dirt contains antidepressant bacteria:* "Dirt Exposure 'Boosts Happiness,'" *BBC News*, last updated April 1, 2007, http://news.bbc.co.uk/2/hi/health /6509781.stm.

39. Bring on the Awe

Page 100, *"A pillar of gold light beamed diagonally from a small hole":* Joe Simpson, *Touching the Void: The True Story of One Man's Miraculous Survival*, rev. ed. (New York: Perennial, 2004), 132–33.

Page 100, *In 2015, researchers tested the behaviors of participants:* Paul K. Piff et al., "Awe, the Small Self, and Prosocial Behavior," *Journal of Personality and Social Psychology* 108, no. 6 (2015): 883–99, https://www.apa.org/pubs/journals/releases/psp -pspi0000018.pdf.

40. Exude Gratitude

Page 102, *Counting one's blessings helps anyone feel better:* Randy A. Sansone and Lori A. Sansone, "Gratitude and Well Being: The Benefits of Appreciation," *Psychiatry* 7, no. 11 (November 7, 2010): 18–22, https://www.ncbi.nlm.nih.gov/pmc/articles /PMC3010965.

Page 102, *Studies affirm that young people who are grateful:* Jeffrey J. Froh, William J.

Sefick, and Robert A. Emmons, "Counting Blessings in Early Adolescents: An Experimental Study of Gratitude and Subjective Well-Being," *Journal of School Psychology* 46 (2008): 213–33, doi:10.1016/j.jsp.2007.03.005.

41. Time Travel

Page 105, *how 1.1 billion people worldwide still live — without electricity:* "Energy Access," International Energy Agency, accessed October 13, 2107, http://www.iea .org/energyaccess.

42. Throw Out the Baby-Safety Catalog

Page 110, *SIDS, for example, is now considered largely preventable:* Andrea Hsu, "Rethinking SIDS: Many Deaths No Longer a Mystery," *NPR*, July 15, 2011, http://www.npr.org/2011/07/15/137859024/rethinking-sids-many-deaths-no -longer-a-mystery.

43. Teach a New Set of Life Skills

Page 111, *To create this list…as well as one by Julie Lythcott-Haims:* As I describe, this chapter's list incorporates ideas by Julie Lythcott-Haims in her book *How to Raise an Adult: Break Free of the Overparenting Trap and Prepare Your Kid for Success* (New York: Henry Holt and Company, 2015), 166–69.

45. Let Kids Make Laws

Page 117, *YouCAN families wrote letters to the editor to support the unprecedented law:* Saul Hubbard, "Goals for Carbon Reduction Become Law in Eugene," *Register-Guard*, July 29, 2014, http://projects.registerguard.com/rg/news/local/31933682-75 /ordinance-climate-eugene-council-cost.html.csp.

Page 118, *YouCAN's campaign has moved to several other cities:* "Grassroots Legal Actions," Our Children's Trust, https://www.ourchildrenstrust.org/grassroots-legal -actions.

48. Take Kids on a Real-Life Hero's Journey

Page 124, *Perhaps the statistics about the alarming — and rising — rates of:* Susanna Schrobsdorff, "There's a Startling Increase in Major Depression Among Teens in the U.S.," *Time*, November 16, 2016, http://time.com/4572593/increase -depression-teens-teenage-mental-health.

49. Re-story Our Future

Page 126, *Also keep in mind a 2013 New School for Social Research study:* Julianne Chiaet, "Novel Finding: Reading Literary Fiction Improves Empathy," *Scientific*

American, October 4, 2013, https://www.scientificamerican.com/article/novel-finding-reading-literary-fiction-improves-empathy.

51. Model Love, Always

Page 131, *mirror neurons move…harmonize human emotional states:* Jonah Lehrer, "The Mirror Neuron Revolution: Explaining What Makes Humans Social," *Scientific American*, accessed October 13, 2017, https://www.scientificamerican.com/article/the-mirror-neuron-revolut.

52. Teach Civics – Not Reading – to Kindergartners

Page 133, *William Deresiewicz calls the "artificial scarcity of educational resources":* William Deresiewicz, "The Neoliberal Arts: How College Sold Its Soul to the Market," *Harper's*, September 2015, https://harpers.org/archive/2015/09/the-neoliberal-arts/8.

53. Empower Kids to Solve Local Problems

Page 135, *EPIC is so successful that it's already been replicated at:* "Sustainable Cities Initiative: University of Oregon," EPIC-Network, accessed October 13, 2017, http://www.epicn.org/university-of-oregon-sustainable-cities-initiative.

55. Let Kids Play with Knives

Page 141, *some pediatricians even write "park prescriptions":* "Kaiser Permanente Pediatrician Writes 'Nature Prescriptions' for Youth," Kaiser Permanente, April 19, 2011, https://share.kaiserpermanente.org/article/kaiser-permanente-pediatrician-writes-nature-prescriptions-for-youth.

Page 141, *empowers children to learn independence and appropriate boundaries:* Roger Mackett, "Unsupervised Children Are More Sociable and More Active, Study Says," *ScienceDaily*, December 18, 2007, https://www.sciencedaily.com/releases/2007/12/071218192030.htm.

56. Fight for Climate Literacy in Schools

Page 142, *I'm sorry to report that according to a 2016 study:* Eric Plutzer et al., "Mixed Messages: How Climate Change Is Taught in America's Public Schools," National Center for Science Education, March 2016, accessed October 1, 2017, https://ncse.com/files/MixedMessages.pdf, 18.

Page 142, *The bad news is that the Heartland Institute — a well funded conservative think-tank:* Angela Fritz, "A Political Organization that Doubts Climate Science Is Sending This Book to 200,000 Teachers," *Washington Post*, March 29, 2017, https://www.washingtonpost.com/news/capital-weather-gang/p/2017/03/29/a-political-organization-that-doubts-climate-science-is-sending-this-book-to-200000-teachers.

Page 142, *In 2016, Portland's school board unanimously passed a first-in-the-United States resolution:* Melissa Hellmann, "Portland Public Schools First to Put Global Climate Justice in Classroom," *Yes! Magazine*, April 6, 2017, http://www.yes magazine.org/planet/portland-public-schools-first-to-put-global-climate-justice -in-classroom-20170406.

Page 143, *Seventeen-year-old Tyler Honn told* Yes! Magazine: Ibid.

57. Unplug the Kids

Page 145, *kids spent an average of* nine hours a day *on-screen:* "The Common Sense Census: Media Use by Tweens and Teens," Common Sense Media, November 3, 2015, https://www.commonsensemedia.org/the-common-sense-census-media-use-by -tweens-and-teens-infographic.

Page 145, *I would do it again to protect their health:* Victoria L. Dunckley, *Reset Your Child's Brain* (Novato, CA: New World Library, 2015), 107.

Page 146, *tech entertainment companies fund many of these studies:* Richard Freed, *Wired Child: Reclaiming Childhood in a Digital Age* (North Charleston, SC: CreateSpace Independent Publishing Platform, 2015), 5.

Page 146, *They're fine. Yours will be, too:* Dunckley, *Reset Your Child's Brain*, 282.

63. Host a Disaster Party

Page 161, *as one did in 2013 in Lac-Mégantic, Quebec, killing forty-seven people:* Associated Press, "Lac-Mégantic Oil Train Disaster Inquiry Finds String of Safety Failings," *The Guardian*, August 19, 2014, https://www.theguardian.com/world /2014/aug/20/lac-megantic-oil-train-disaster-inquiry-finds-string-of-safety -failings.

Page 161, *"Our operating assumption is that everything west":* Kathryn Schulz, "The Really Big One," *New Yorker*, July 20, 2015, https://www.newyorker.com /magazine/2015/07/20/the-really-big-one.

64. Kick Out the Climate Blues

Page 164, *A 2015 Oxford University study explains why: When we dance:* Bronwyn Tarr et al., "Synchrony and Exertion During Dance Independently Raise Pain Threshold and Encourage Social Bonding," *Biology Letters* 11, no. 10 (October 28, 2015), http://rsbl.royalsocietypublishing.org/content/11/10/20150767.

67. Tame Your Tongue

Page 173, *Nonviolent communication (NVC) improves communication:* "What Is NVC," Center for Nonviolent Communication, accessed October 13, 2017, https://www .cnvc.org/about/what-is-nvc.html.

68. Let Your Farmer Feed You

Page 176, *scientists have found dirt makes us happier:* "Dirt Exposure 'Boosts Happiness,'" *BBC News*, last updated April 1, 2007, http://news.bbc.co.uk/2/hi/health/6509781.stm.

69. Hound Representatives

Page 178, Indivisible *is an easy-to-read, twenty-six-page document:* "Indivisible: A Practical Guide for Resisting the Trump Agenda," Indivisible, updated March 9, 2017, https://www.indivisible.org/guide.

70. Demand Clean Energy – While Filling Your Tank

Page 181, *The footprint of an EV is half — or less:* Nealer, Reichmuth, and Anair, "Cleaner Cars."

71. Escort Big Oil from the Museum – and Off Campus

Page 183, *Billionaire oil baron and antigovernment zealot David Koch paid the museum:* Joe Romm, "Smithsonian Stands By Wildly Misleading Climate Change Exhibit Paid for by Kochs," ThinkProgress, March 23, 2015, https://thinkprogress.org/smithsonian-stands-by-wildly-misleading-climate-change-exhibit-paid-for-by-kochs-bd3105ef354b.

Page 183, *"How do you think your body will evolve?":* Ibid.

Page 183, *To this end, the Kochs paid nearly $900 million in the 2016 election:* Alan Yuhas, "David Koch Steps Down from Board of New York Science Museum," *The Guardian*, January 21, 2016, https://www.theguardian.com/us-news/2016/jan/21/david-koch-american-museum-of-natural-history-climate-change-fossil-fuel-money.

Page 184, *In 2015, in an unprecedented open letter to science and natural history:* "An Open Letter to Museums from Members of the Scientific Community," Natural History Museum, March 24, 2015, http://thenaturalhistorymuseum.org/open-letter-to-museums-from-scientists.

Page 184, *Koch resigned from the board:* Yuhas, "David Koch Steps Down."

Page 184, *more institutions worldwide are divesting daily:* Katherine Bagley, "Science Museums Cutting Financial Ties to Fossil Fuel Industry," *InsideClimate News*, November 16, 2015, https://insideclimatenews.org/news/15112015/science-museums-cut-financial-ties-fossil-fuel-industry-field-museum-california-academy-sciences.

Page 184, *Remember, the tobacco industry once pulled the same junk-science stunts:* Graham Readfearn, "Doubt Over Climate Science Is a Product with an Industry Behind It," *The Guardian*, March 5, 2015, https://www.theguardian.com/environment/planet-oz/2015/mar/05/doubt-over-climate-science-is-a-product-with-an-industry-behind-it.

73. Become a Victory Speaker

Page 187, *He and Congress raised taxes, passed laws like the:* "H.R. 1776: A Bill Further to Promote the Defense of the United States and for Other Purposes" (Lend-Lease Bill), January 10, 1941, National Archives, https://www.archives.gov /historical-docs/todays-doc/?dod-date=110.

Page 187, *By 1942, industries hummed along in a government-coordinated:* Frank N. Schubert, "Mobilization: The U.S. in World War II, The 50th Anniversary," brochure, U.S. Army Center of Military History, accessed October 11, 2017, https://history.army.mil/documents/mobpam.htm.

Page 187, *victory speakers — ordinary citizens who volunteered to give daily:* "Preparation for Victory Speaking for Use by War Speaker's Bureau," accessed July 1, 2017, Oregon Secretary of State archives, http://sos.oregon.gov/archives/exhibits/ww2 /Documents/life-speak3.pdf. See also "An Arsenal of Words: Victory Speakers Spread the Word," Oregon Secretary of State archives, accessed November 28, 2017, http://sos.oregon.gov/archives/exhibits/ww2/Pages/life-speakers.aspx.

Page 188, *We have no Winston Churchill reminding us that, "without victory":* "'Victory at All Costs,'" *The Guardian*, May 14, 1940, https://www.theguardian.com /century/1940-1949/Story/0,,127386,00.html.

74. Trumpet the Pope

Page 189, *In his 2015 papal letter addressed to "every person living on this planet":* Pope Francis, *Laudato Si: On Care for Our Common Home*, encyclical letter (May 24, 2015), sections 3 and 114, http://w2.vatican.va/content/dam/francesco/pdf /encyclicals/documents/papa-francesco_20150524_enciclica-laudato-si_en.pdf.

Page 189, *"Doomsday predictions can...concealing their symptoms":* Ibid., sections 161, 14, and 26.

Page 189, *He challenged them boldly to "do unto others":* Nahal Toosi, "Pope to Congress: 'Do Unto Others...'", *Politico*, September 24, 2015, http://www.politico .com/story/2015/09/pope-francis-addresses-congress-214008.

Page 189, *Many Republicans mocked him, but the majority:* Suzanne Goldenberg and Sabrina Siddiqui, "Jeb Bush Joins Republican Backlash Against Pope on Climate Change," *The Guardian*, June 17, 2015, https://www.theguardian.com/us-news /2015/jun/17/jeb-bush-joins-republican-backlash-pope-climate-change.

Page 189, *Within weeks, 1.8 million people signed several:* ACT Alliance press release, "Nearly 1.8 Million People Demand Climate Action," November 28, 2015, http://actalliance.org/act-news/nearly-1-8-million-people-demand-climate-action.

Page 189, *The Dalai Lama called global warming a "problem which":* Associated Press, "Dalai Lama Says Strong Action on Climate Change Is a Human Responsibility," *The Guardian*, October 20, 2015, https://www.theguardian.com/environment /2015/oct/20/dalai-lama-says-strong-action-on-climate-change-is-a-human -responsibility.

Page 189, *Archbishop Desmond Tutu went further. He said, "Time"*: "Desmond Tutu: Climate Change Is the Human Rights Challenge of Our Time," Desmond Tutu Peace Foundation, March 28, 2017, http://www.tutufoundationusa.org /2017/03/28/desmond-tutu-climate-change-human-rights-challenge-time.

Page 190, *young people copied lines promoting care of the Earth:* Nature's Calling, DoSomething.org, accessed January 29, 2017, https://www.dosomething.org /us/campaigns/natures-calling.

75. Shout from the Solar Roller Coaster

Page 192, *even processing it in a deeper-thinking part of my brain:* Alina Tugend, "Praise Is Fleeting, but Brickbats We Recall," *New York Times*, March 23, 2012, http://www.nytimes.com/2012/03/24/your-money/why-people-remember -negative-events-more-than-positive-ones.html.

Page 192, *or how American solar employs more people than:* Niall McCarthy, "Solar Employs More People in U.S. Electricity Generation than Oil, Coal, and Gas Combined," *Forbes*, January 25, 2017, https://www.forbes.com/sites/niallmccarthy /2017/01/25/u-s-solar-energy-employs-more-people-than-oil-coal-and-gas -combined-infographic/#73c3d70c2800.

Page 193, *Denmark's biggest energy company abandoned fossil fuels:* Akshat Rathi, "Denmark's Biggest Energy Company Is Completely Abandoning Fossil Fuels," *Quartz*, October 3, 2017, https://qz.com/1092873/dong-energy-is-abandoning -fossil-fuels-and-changing-its-name-to-orsted-energy.

Page 193, *TransCanada canceled a major tar sands pipeline project:* Nicholas Kusnetz, "Major Tar Sands Oil Pipeline Cancelled, Dealing Blow to Canada's Export Hopes," *InsideClimate News*, October 5, 2017, https://insideclimatenews.org /news/05102017/transcanada-tar-sands-oil-pipeline-canada-energy-east -climate-emissions.

Page 193, *China is building the world's first smog-eating "Forest City!":* Allyssia Alleyne, "China Unveils Plans for World's First Pollution-Eating 'Forest City,'" *CNN*, July 20, 2017, http://www.cnn.com/style/article/china-liuzhou-forest-city/index .html. See also Simon Brandon, "China Just Switched On the World's Largest Floating Solar Power Plant," *World Economic Forum*, June 2, 2017, https://www .weforum.org/agenda/2017/06/china-worlds-largest-floating-solar-power.

Page 193, *France banned new oil and gas exploration:* Lizzie Dearden, "France to Ban All New Oil and Gas Exploration in Renewable Energy Drive," *The Independent*, June 24, 2017, http://www.independent.co.uk/news/world/europe/france-ban-new -oil-gas-exploration-stop-granting-licences-macron-hulot-renewable-energy -drive-a7806161.html.

Page 193, *Copenhagen has more bikes than cars:* Danielle Muoio, "Copenhagen Now Has More Bikes than Cars — Here Are the 5 Most Bike-Friendly Cities in the World," *Business Insider*, March 6, 2017, http://www.businessinsider.com/copenhagen-has -more-bikes-than-cars-2017-3/#5-eindhoven-netherlands-1.

Page 193, *Nevada and Washington passed laws:* "New Washington Solar Bill Is a Move in

the Right Direction," NW Energy Coalition, July 5, 2017, http://www.nwenergy
.org/news/new-washington-solar-bill-is-a-move-in-the-right-direction. See also
Jeff Brady, "Solar Firms Plan to Return to Nevada After New Law Restores Incentives," *NPR*, June 7, 2017, http://www.npr.org/sections/thetwo-way/2017
/06/07/531952407/solar-firms-plan-to-return-to-nevada-after-new-law
-restores-incentives.

Page 193, *In New Zealand and Ecuador, rivers now have legal rights:* Mihnea Tanasescu,
"Rivers Get Human Rights: They Can Sue to Protect Themselves," *Scientific
American*, June 19, 2017, https://www.scientificamerican.com/article/rivers
-get-human-rights-they-can-sue-to-protect-themselves.

Page 193, *California's Great America amusement park is powered by* 100 percent: "California's Great America Purchases 100% Renewable Energy from Silicon Valley Power
to Cover Electricity Use," *Business Wire*, April 21, 2017, http://www.businesswire
.com/news/home/20170421005145/en/California%E2%80%99s-Great-America
-Purchases-100-Renewable-Energy.

76. Push Your City to Get Ready

Page 194, *A month later, an algae bloom covered much of Lake Erie:* Jugal K. Patel and
Yuliya Parshina-Kottas, "Miles of Algae Covering Lake Erie," *New York Times*,
October 3, 2017, https://www.nytimes.com/interactive/2017/10/03/science
/earth/lake-erie.html.

Page 194, *blazes swept California, incinerating almost nine thousand homes:* Alex Dobu-
zinskis, "Death Toll from California Blazes Rises to 43, After Teen Dies," *Reuters*,
October 30, 2017, http://www.reuters.com/article/us-california-fire/death-toll
-from-california-blazes-rises-to-43-after-teen-dies-idUSKBN1D004Z.

Page 194, *The South is scheduled to bake:* Solomon Hsiang et al., "Estimating Economic
Damage from Climate Change in the United States," *Science* 356, no. 6345 (June
30, 2017), http://science.sciencemag.org/content/356/6345/1362.

Page 194, *and coastal cities…regularly flood and become partly submerged:* William
V. Sweet et al., "Global and Regional Sea Level Rise Scenarios for the United
States," NOAA, January 2017, https://tidesandcurrents.noaa.gov/publications
/techrpt83_Global_and_Regional_SLR_Scenarios_for_the_US_final.pdf.

Page 194, *The Mile-High City is working to become a healthy:* "2020 Sustainability
Goals," Denver Office of Sustainability, accessed October 13, 2017, https://www
.denvergov.org/content/denvergov/en/office-of-sustainability/2020
-sustainability-goals.html.

Page 195, *Or Portland, which has invested…to its students:* "Climate Action Plan Progress Report," City of Portland, Oregon, and Multnomah County, April 2017,
https://www.portlandoregon.gov/bps/article/636700. See also Alison Flood,
"Portland Schools Ditch Textbooks That Question Climate Change," *The Guardian*, May 24, 2016, https://www.theguardian.com/books/2016/may/24/portland
-schools-ditch-textbooks-that-question-climate-change.

Page 195, *Or Chicago, helping neighborhoods build the community:* Erin Brodwin, "The

Best US Cities to Live in to Escape the Worst Effects of Climate Change," *Business Insider*, September 24, 2017, http://www.businessinsider.com/best-us-cities-escape-climate-change-2017-9.

Page 195, *Read the* Atlantic *article "The American South Will Bear":* Robinson Meyer, "The American South Will Bear the Worst of Climate Change's Costs," *Atlantic*, June 2017, http://www.theatlantic.com/science/archive/2017/06/global-warming-american-south/532200.

77. Connect Your Work World to the Real World

Page 196, *Sutter's solution was so unexpected that it made the front page:* Jess Bidgood, "Charges Dropped Against Climate Activists," *New York Times*, September 8, 2014, https://www.nytimes.com/2014/09/09/us/charges-dropped-against-climate-activists.html.

Page 196, *Sutter said … "but was also made with our concerns":* Jon Queally, "'They're Right': Citing Climate, Prosecutor Drops Charges Against Coal Blockaders," *Common Dreams*, September 9, 2014, https://www.commondreams.org/news/2014/09/09/theyre-right-citing-climate-prosecutor-drops-charges-against-coal-blockaders.

Page 196, *He later told* Democracy Now!*: "To the extent that":* "Exclusive: DA Joins the Climate Activists He Declined to Prosecute, Citing Danger of Global Warming," *Democracy Now!*, September 10, 2014, https://www.democracynow.org/2014/9/10/exclusive_da_joins_the_2_climate.

78. Know Exactly What You're Voting For

Page 198, *I join the majority of both Democrats and Republicans calling for electoral:* "Beyond Distrust: How Americans View Their Government: 6. Perceptions of Elected Officials and the Role of Money in Politics," Pew Research Center, November 23, 2015, http://www.people-press.org/2015/11/23/6-perceptions-of-elected-officials-and-the-role-of-money-in-politics.

Page 198, *on climate policy alone, the Democratic and Republican Party platforms:* "Building a Clean Energy Economy," 2016 Democratic Party Platform, Democrats.org, accessed August 2, 2017, https://www.democrats.org/party-platform#environment; and "America's Natural Resources: Agriculture, Energy, and the Environment," 2016 Republican Party Platform, GOP.com, accessed August 2, 2017, https://gop.com/platform/americas-natural-resources.

Page 198, *President Nixon signed the Clean Air Act:* "40th Anniversary of the Clean Air Act," EPA, September 14, 2010, https://www.epa.gov/clean-air-act-overview/40th-anniversary-clean-air-act.

Page 198, *Here, take a look at this quick comparison:* Phil McKenna, "GOP and Democratic Platforms Highlight Stark Differences on Energy and Climate," *InsideClimate News*, July 26, 2016, https://insideclimatenews.org/news/26072016

/democrat-republican-party-platforms-energy-climate-change-hillary-clinton
-donald-trump.

Page 198, *Calls climate change a "hoax"*: Maya Rhodan, "Donald Trump Calls Climate
Change a Hoax, but Worries It Could Hurt His Golf Course," *Time*, May 23,
2016, http://time.com/4345367/donald-trump-climate-change-golf-course.

Page 198, *Unveils plans to become "the clean energy superpower"*: Sabrina Siddiqui,
"Hillary Clinton Unveils Her Plan to Make US 'Clean Energy Superpower',"
The Guardian, September 23, 2015, https://www.theguardian.com/environment
/2015/sep/24/hillary-clinton-unveils-her-masterplan-to-make-us-clean-energy
-superpower.

Page 199, *Calls EPA regulations a "disaster"*: Tessa Berenson, "Donald Trump Says
EPA Water Regulations Are Hurting His Hair," *Time*, December 12, 2015,
http://time.com/4146809/donald-trump-epa-water-hair.

Page 199, *Cut energy programs…studying oceans and climate*: Michael Greshko, Laura
Parker, and Brian Clark Howard, "A Running List of How Trump Is Changing
the Environment," *National Geographic*, last updated December 18, 2017,
http://news.nationalgeographic.com/2017/03/how-trump-is-changing-science
-environment.

Page 200, *The GOP platform…accuses Democrats of being "extremists"*: "America's
Natural Resources," 2016 Republican Party Platform.

Page 200, *Read the* National Geographic *article "A Running List"*: Michael Greshko,
Laura Parker, and Brian Clark Howard, "A Running List of How Trump Is
Changing the Environment," *National Geographic*, last updated December 18 2017,
http://news.nationalgeographic.com/2017/03/how-trump-is-changing-science
-environment.

79. Bury Your Neighbor's Chicken

Page 201, *Jacqueline Woodson tells a story, in her beautiful free-verse*: Jacqueline
Woodson, *Brown Girl Dreaming* (New York: Nancy Paulsen Books, 2014), 80, 81.

80. Rescue Food

Page 203, *that* one-third *of all food produced worldwide is thrown out*: Hawken, *Drawdown*, 42.

Page 203, *In the United States, it's 40 percent*: Dana Gunders, "Wasted: How America Is
Losing Up to 40 Percent of Its Food from Farm to Fork to Landfill," Natural
Resources Defense Council, August 2012, 4, https://www.nrdc.org/sites/default
/files/wasted-food-IP.pdf.

Page 203, *cut down on 8 percent of the world's human-caused emissions*: Hawken, *Drawdown*, 42.

Page 203, *Entire crops often rot in the field*: Gunders, "Wasted," 7, 8.

Page 203, *Everywhere along the food-supply chain, workers swoop in*: Elizabeth Royte,
"How 'Ugly' Fruits and Vegetables Can Help Solve World Hunger," *National*

Geographic, March 2016, https://www.nationalgeographic.com/magazine/2016/03/global-food-waste-statistics.

Page 203, *But the bulk of food waste — two-thirds, to be precise — happens in:* Emily Broad Leib et al., "Consumer Perceptions of Date Labels: National Survey," Harvard Food Law and Policy Clinic, May 2016, https://www.chlpi.org/wp-content/uploads/2013/12/consumer-perceptions-on-date-labels_May-2016.pdf. For a summary of the full report, see "FLPC Releases Report on Consumer Perceptions of Food Date Labels," Center for Health Law and Policy Innovation, Harvard Law School, May 11, 2016, https://www.chlpi.org/flpc-releases-report-on-consumer-perceptions-of-food-date-labels.

Page 203, *though it is true that we're far less skilled at kitchen efficiency:* Hawken, *Drawdown*, 42, 43.

Page 203, *In fact, a 2016 study on food date labels concluded, "With only":* Leib, "Consumer Perceptions."

Page 204, *If possible, use glass. (Plastic containers leach chemicals):* Joseph Mercola, "Practical Options to Store Your Food without Contaminating Them with Plastics," November 16, 2011, https://articles.mercola.com/sites/articles/archive/2011/11/16/practical-options-to-store-your-food-without-contaminating-them-with-plastics.aspx.

81. Kick Compost to the Curb

Page 205, *Soon after, I attended her keynote speech to Northwest waste specialists:* Details in this and the next paragraph are from Brenda Platt, "Composting Comes Full Circle," Institute for Local Self-Reliance, PowerPoint presentation, November 16, 2016, https://ilsr.org/brenda-platt-keynote-worc-2016.

Page 206, *Moreover, when compost is spread on the ground, it pulls carbon:* Rebecca Ryals and Whendee L. Silver, "Can Land Management Enhance Soil Carbon Sequestration?" Marin Carbon Project, accessed October 13, 2017, http://www.marincarbonproject.org/science/land-management-carbon-sequestration.

Page 206, *we'd reduce carbon in the atmosphere to the safe concentration:* Nathanael Johnson, "Just Add Compost: How to Turn Your Grassland Ranch into a Carbon Sink," *Grist*, January 16, 2014, http://grist.org/climate-energy/just-add-compost-how-to-turn-your-grassland-ranch-into-a-carbon-sink.

Page 206, *Happily, several cities and states are:* Emily S. Rueb, "How New York Is Turning Food Waste into Compost and Gas," *New York Times*, June 2, 2017, https://www.nytimes.com/2017/06/02/nyregion/compost-organic-recycling-new-york-city.html.

82. Do the New Math

Page 207, *Because there's one equation that clarifies this moment:* Bill McKibben, *Oil and Honey: The Education of an Unlikely Activist* (New York: Henry Holt and Company, 2013), 141–50.

Page 207, *to raise Earth's temperature .08 degrees Celsius:* McKibben, *Oil and Honey*, 143.

Page 208, *That's why industry denies the above-mentioned math:* Readfearn, "Doubt Over Climate Science."

Page 208, *They told the huge crowds: "It's a lot like nuclear overkill":* McKibben, *Oil and Honey*, 148.

Page 208, *they left youth-led divestment campaigns — 252 by the tour's end:* McKibben, *Oil and Honey*, 237.

Page 208, *the* New York Times *called the "vanguard of a new national movement":* McKibben, *Oil and Honey*, 236.

83. Divest. Get Everyone To.

Page 209, *By holding a mock wedding, University of Oregon (UO) students:* Andrew Theen, "University of Oregon Students Stage Mock Wedding between UO Foundation and Fossil Fuel Industry," *The Oregonian*, April 22, 2016, http://www.oregonlive.com/education/index.ssf/2016/04/university_of_oregon_students_5.html.

Page 209, *Five months later, UO announced its plans to divest:* Diane Dietz, "UO Plans to Divest Fossil Fuel Holdings," *Register-Guard*, September 13, 2016, http://projects.registerguard.com/rg/news/local/34790290-75/uo-foundation-is-on-the-road-to-divesting-in-fossil-fuels-website-says.html.csp.

Page 209, *Countless individuals and 850-plus institutions have divested:* Damian Carrington, "Fossil Fuel Divestment Funds Double to $5tn in a Year," *The Guardian*, December 12, 2016, https://www.theguardian.com/environment/2016/dec/12/fossil-fuel-divestment-funds-double-5tn-in-a-year.

Page 209, *"Just as we argued in the 1980s that those who conducted business":* "Desmond Tutu: Climate Change Is the Human Rights Challenge of Our Time," Desmond Tutu Peace Foundation, March 28, 2017, http://www.tutufoundationusa.org/2017/03/28/desmond-tutu-climate-change-human-rights-challenge-time.

Page 209, *Energy Transfer Partners…has sued Greenpeace and other groups:* "It's Time to Boycott the Banks Funding Climate Disaster," 350.org, October 10, 2017, https://350.org/boycott-banks-funding-climate-disaster.

Page 209, *That's because it's not only colleges and churches dumping the dirty stuff:* "Divestment Commitments," 350.org, accessed October 31, 2017, https://gofossilfree.org/divestment/commitments.

Page 210, *investors are realizing they stand to lose trillions:* "The $2 Trillion Stranded Assets Danger Zone: How Fossil Fuel Firms Risk Destroying Investor Returns," Carbon Tracker Initiative, November 24, 2015, https://www.carbontracker.org/reports/stranded-assets-danger-zone.

Page 210, *World Bank president Jim Yong Kim declared, "Sooner rather than":* "World Bank Group President Jim Yong Kim Remarks at Davos Press Conference," World Bank, January 23, 2014, http://www.worldbank.org/en/news/speech/2014/01/23/world-bank-group-president-jim-yong-kim-remarks-at-davos-press-conference.

Page 210, *He even told the* Guardian, *when calling for a carbon tax:* Larry Elliott, "Scrap

Fossil Fuel Subsidies Now and Bring In Carbon Tax, Says World Bank Chief,"
The Guardian, April 13, 2015, https://www.theguardian.com/environment/2015
/apr/13/fossil-fuel-subsidies-say-burn-more-carbon-world-bank-president.
Page 210, *at their 2017 meeting, they passed a resolution:* Steven Mufson, "Financial Firms
Lead Shareholder Rebellion Against ExxonMobil Climate Change Policies,"
Washington Post, May 31, 2017, https://www.washingtonpost.com/news
/energy-environment/wp/2017/05/31/exxonmobil-is-trying-to-fend-off-a
-shareholder-rebellion-over-climate-change.

85. Empower Women and Girls Everywhere

Page 215, *Literacy is the number-one factor:* Hawken, *Drawdown*, 81, 82.

87. Level Up the Antibullying Campaigns

Page 220, *These campaigns effectively stall critical conversations:* Readfearn, "Doubt
Over Climate Science."
Page 220, *"In fourth grade,"* one member said in a meeting with lawmakers: "DC Bully
Busters — Working to End Bullying in Politics," video, DC Bully Busters, ac-
cessed October 12, 2017, http://dcbullybusters.com.

88. Creatively Disrupt

Page 221, *pie-in-the-face political protests:* "Who Got a Pie in the Face?" *CBS News*,
accessed October 13, 2017, https://www.cbsnews.com/pictures/who-got-a-pie
-in-the-face/3.
Page 222, *Gag orders forbid some state employees:* "The Truth about Florida's Attempt to
Censor Climate Change," Union of Concerned Scientists, April 2015, http://www
.ucsusa.org/publications/got-science/2015/got-science-april-2015#.WejdJiMrInW.
Page 222, *President Trump barred EPA and Department of Agriculture scientists:* Angela
Chen, "Trump Silences Government Scientists with Gag Orders," *The Verge*,
January 24, 2017, https://www.theverge.com/2017/1/24/14372940/trump-gag
-order-epa-environmental-protection-agency-health-agriculture.
Page 222, *As Bill McKibben writes, "Winning slowly":* Bill McKibben, "A World at
War," *New Republic*, August 15, 2016, https://newrepublic.com/article/135684
/declare-war-climate-change-mobilize-wwii.

89. Support Grassroots Groups

Page 225, *Unbelievably, many even take money from or forge "partnerships":* Naomi Klein,
This Changes Everything: Capitalism vs. the Climate (New York: Simon and Schus-
ter, 2014), 196.
Page 225, *Keeping coal, oil, and gas in the ground? (Nope.):* "Stand Up for Real Climate
Action," the Nature Conservancy, accessed October 13, 2017, https://support
.nature.org/site/Advocacy?cmd=display&page=UserAction&id=182.

Page 225, *Do they "partner" with polluters Dow Chemical:* "Working with Companies: Companies We Work With," The Nature Conservancy, accessed October 13, 2017, https://www.nature.org/about-us/working-with-companies/companies -we-work-with/index.htm.

Page 225, *Do they ask you to calculate your carbon footprint?:* "Carbon Calculator," The Nature Conservancy, accessed October 13, 2017, https://www.nature.org/green living/carboncalculator/index.htm?redirect=https-301&ref=www.nature.org.

Page 225, *Count the major polluters on a Big Green business council:* "Working with Companies: Business Council," The Nature Conservancy, accessed October 13, 2017, https://www.nature.org/about-us/working-with-companies/businesscouncil /ilc-main-content.xml.

Page 225, *reported a 2012 study by the National Committee for Responsive Philanthropy:* Sarah Hansen, "Cultivating the Grassroots," National Committee for Responsive Philanthropy (February 23, 2012), 1, https://www.ncrp.org/publication /cultivating-the-grassroots.

Page 225, *If green philanthropists don't… "affected by environmental ills":* Ibid., 15, 32–33.

Page 226, *most Big Greens actually have no policy against investing:* Klein, *This Changes Everything*, 197.

Page 226, *and as Gandhi said, "A body of determined":* Hansen, "Cultivating the Grassroots," 11.

90. Interrupt the Mars and Geoengineering Fantasies

Page 227, *For decades, the fossil-fuel industry has lied:* Readfearn, "Doubt Over Climate Science."

Page 227, *McCaughrean tweeted, "I'm less concerned about making humans":* Hannah Devlin, "Life on Mars: Elon Musk Reveals Details of His Colonisation Vision," June 16, 2017, *The Guardian*, https://www.theguardian.com/science/2017 /jun/16/life-on-mars-elon-musk-reveals-details-of-his-colonisation-vision.

Page 227, *The most popular version involves lowering Earth's temperature:* Klein, *This Changes Everything*, 256.

Page 228, *And unjust because climatologists predict it will further decrease rainfall:* Ibid., 260.

Page 228, *According to MIT Technology Review, they hope to conduct:* James Temple, "Harvard Scientists Moving Ahead on Plans for Atmospheric Geoengineering Experiments," *MIT Technology Review*, March 24, 2017, https://www .technologyreview.com/s/603974/harvard-scientists-moving-ahead-on-plans -for-atmospheric-geoengineering-experiments.

Page 228, *"The solution to pollution is…pollution?":* Klein, *This Changes Everything*, 256.

91. Inoculate Everyone

Page 230, *He calls public schools…nuclear waste over our oceans:* David Sarasohn, "Art Robinson Didn't Really Mean Those Things. Did He?" *The Oregonian*, October

23, 2010, http://www.oregonlive.com/news/oregonian/david_sarasohn/index .ssf/2010/10/art_robinson_didnt_really_mean.html.

Page 230, *Indeed, the petition boasts 31,487 signatures:* Global Warming Petition Project, accessed July 20, 2017, http://www.petitionproject.org.

Page 230, *Signatories have included Charles Darwin and characters from the TV show:* "Oregon Petition," DeSmog blog, accessed October 31, 2017, https://www .desmogblog.com/oregon-petition.

Page 230, *And only thirty-nine signatories even claim to be climatologists:* Global Warming Petition Project.

Page 230, *In just two easy steps, we can use this to "inoculate" people:* Sander van der Linden et al., "Inoculating the Public against Misinformation about Climate Change," *Global Challenges* (January 23, 2017), doi:10.1002/gch2.201600008, http://climatecommunication.yale.edu/wp-content/uploads/2017/01/Inoculation -article-Global-Challenges.pdf.

Page 231, *that, for example, one of the Spice Girls "signed" twice:* "Oregon Petition," DeSmog.

Page 231, *Just those two steps, according to a University of Cambridge study:* van der Linden, "Inoculating the Public."

Page 231, *According to a study by University of Queensland's John Cook:* John Cook, Stephan Lewandowsky, and Ullrich K. H. Ecker, "Neutralizing Misinformation through Inoculation: Exposing Misleading Argumentation Techniques Reduces Their Influence," PLOS ONE 12, no. 5 (May 5, 2017), https://doi.org/10.1371 /journal.pone.0175799.

Page 231, *students are intrigued by this dramatic story:* Michelle Nijhuis, "Scientists Are Testing a 'Vaccine' Against Climate Change Denial," *Vox*, May 31, 2017, https:// www.vox.com/science-and-health/2017/5/31/15713838/inoculation -climate-change-denial.

92. Battle to Win

Page 232, *Fossil-fuel infestors...want our public lands:* "America's Natural Resources: Agriculture, Energy, and the Environment," Republican Party Platform, GOP .com, accessed October 9, 2017, https://gop.com/platform/americas-natural -resources.

Page 233, *ExxonMobil has had forty years — since learning that its product:* Geoffrey Supran and Naomi Oreskes, "ExxonMobil's Climate Change Communications (1977–2014)," *Environmental Research Letters* 12, no. 8 (August 23, 2017), http://iopscience.iop.org/article/10.1088/1748-9326/aa815f.

93. Break Human Laws (to Obey Nature's Laws)

Page 234, *Dr. James Hansen called "the fuse to the biggest carbon bomb":* Elizabeth McGowan, "NASA's Hansen Explains Decision to Join Keystone Pipeline Protests,"

Reuters, August 29, 2011, https://www.reuters.com/article/idUS2575908057 20110829.

Page 234, *But in 2017, the Trump administration…burn to their hearts' content:* Brian Naylor, "Trump Gives Green Light to Keystone, Dakota Access Pipelines," January 24, 2017, https://www.npr.org/2017/01/24/511402501/trump-to-give -green-light-to-keystone-dakota-access-pipelines. See also "President Donald J. Trump Proclaims October 2017 as National Energy Awareness Month," White House press release, October 12, 2017, https://www.whitehouse.gov/presidential -actions/president-donald-j-trump-proclaims-october-2017-national-energy -awareness-month.

Page 234, *Today, nonviolent civil disobedience has a long tradition:* "Honoring Martin Luther King, Jr: Five Examples of Nonviolent, Civil Disobedience Worldwide," School of International Service, American University, January 20, 2014, https://ironline.american.edu/honoring-martin-luther-king-jr.

Page 235, *Tim DeChristopher, who disrupted an oil and gas lease auction:* Brandon Loomis "DeChristopher Sentenced to Prison, 26 Protesters Arrested," *Salt Lake Tribune*, July 27, 2011, http://archive.sltrib.com/article.php?id=52263987&itype=CMSID.

Page 235, *Dr. James Hansen, who locked himself to the White House fence:* "48 Arrested in Civil Disobedience at White House to Stop Keystone XL Pipeline and Push Obama on Climate Action," Climate Science & Policy Watch, February 13, 2013, http://www.climatesciencewatch.org/2013/02/13/48-arrested-in-kxl-civil -disobedience-at-white-house.

Page 235, *Sandra Steingraber, who blocked a fracked gas project:* "Seneca 12 Block Gas Facility; Sandra Steingraber Arrested," *Orion Magazine*, March 19, 2013, https://orionmagazine.org/2013/03/seneca-12-block-gas-facility-sandra -steingraber-arrested.

Page 235, *Iyuskin American Horse, who locked himself to a digger:* Deirdre Fulton, "'World Watching' as Tribal Members Put Bodies in Path of Dakota Pipeline," *Common Dreams*, September 1, 2016, https://www.commondreams.org/news /2016/09/01/world-watching-tribal-members-put-bodies-path-dakota-pipeline.

Page 235, *In 2017, twenty states proposed or passed laws suppressing:* Jake Johnson, "Since Trump's Election, 20 States Have Moved to Criminalize Dissent," *Common Dreams*, June 20, 2017, https://www.commondreams.org/news/2017/06/20 /trumps-election-20-states-have-moved-criminalize-dissent.

Page 235, *Yet, one out of six Americans reports being willing to risk arrest:* Anthony Leiserowitz et al., "Americans' Actions to Limit Global Warming, November 2013," Yale Program on Climate Change Communication, February 19, 2014, http://climatecommunication.yale.edu/publications/americans-actions-to -limit-global-warming-november-2013.

Page 235, *"At this point of unimaginable threats on the horizon":* Bryan Farrell, "Tim DeChristopher: This Is What Hope Looks Like," Waging Nonviolence, July 26, 2011, https://wagingnonviolence.org/feature/tim-dechristopher-this-is-what -hope-looks-like.

94. Make Polluters Pay

Page 236, *As Bill McKibben writes, "The perverse logic of capitalism":* Greg Jobin-Leeds and AgitArte, *When We Fight, We Win!* (New York: The New Press, 2016), 139.

Page 236, *In fact, only* one hundred companies *are responsible for 71 percent:* Paul Griffin, "The Carbon Majors Database," CDP Worldwide, July 2017, https://b8f65cb373b1b7b15feb-c70d8ead6ced55ob4d987d7c03fcdd1d.ssl.cf3 .rackcdn.com/cms/reports/documents/000/002/327/original/Carbon-Majors -Report-2017.pdf.

Page 236, *one International Monetary Fund study that tallied environmental:* David Coady et al., "How Large Are Global Energy Subsidies?" International Monetary Fund Working Paper (May 2015), 5, https://www.imf.org/external/pubs/ft/wp/2015 /wp15105.pdf.

Page 237, *British Columbia's tax slashed carbon emissions by 16 percent:* Diane Toomey, "How British Columbia Gained by Putting a Price on Carbon," *Yale Environment 360*, April 30, 2015, http://e360.yale.edu/features/how_british_columbia _gained_by_putting_a_price_on_carbon.

Page 237, *Ugandan activist Bernadette Kodili Chandia argues in the film:* Andy Bichlbaum, Mike Bonanno, and Laura Nix, *The Yes Men Are Revolting* (2014), DVD.

Page 237, *one study predicts that nine out of ten African farmers won't be:* Louis Fox, "The Story of Cap and Trade" (San Francisco and Boston: Free Range Studios, 2009), film, https://storyofstuff.org/movies/story-of-cap-and-trade.

Page 237, *The Green Climate Fund, an international fund to:* "Who We Are: About the Fund," Green Climate Fund, accessed November 13, 2017, http://www.green climate.fund/who-we-are/about-the-fund.

95. Link Eco-Justice to All Justice

Page 238, *By September, the US Army Corps of Engineers had rejected:* Amy Dalrymple, "Pipeline Route Plan First Called for Crossing North of Bismarck," *Bismarck Tribune*, August 18, 2016, http://bismarcktribune.com/news/state-and-regional /pipeline-route-plan-first-called-for-crossing-north-of-bismarck/article _64d053e4-8a1a-5198-a1dd-498d386c933c.html.

Page 238, *One involves the trampling of indigenous rights and treaties:* "Indigenous Peoples and Industrial Corporations," United Nations Permanent Forum on Indigenous Issues, accessed November 12, 2017, http://www.un.org/en/events /indigenousday/pdf/Indigenous_Industry_Eng.pdf.

Page 238, *According to the NAACP, "Race — even more than class":* "Environmental & Climate Justice," NAACP, accessed October 13, 2017, http://www.naacp.org /issues/environmental-justice.

Page 238, *Industrial neighbors create toxic air, which disproportionately hurts children:* "Asthma & Outdoor Air Pollution in the Latino Community," Moms Clean Air Force, accessed January 1, 2018, http://www.momscleanairforce.org /asthma-outdoor-air. See also "Asthma and African Americans," US Department of Health and Human Services, Office of Minority Health, accessed October 13, 2017, https://minorityhealth.hhs.gov/omh/browse.aspx?lvl=4&lvlid=15.

Page 238, *Poor communities are…struggle more to rebound:* "Hurricane Harvey Hit Low-Income Communities Hardest," *Nexus Media*, September 1, 2017, https://thinkprogress.org/hurricane-harvey-hit-low-income-communities -hardest-6d13506b7e60.

Page 238, *In the developing world, millions of people suffer:* "2016: The State of Food and Agriculture," Food and Agriculture Organization of the United Nations (2016), xi, xii, http://www.fao.org/3/a-i6030e.pdf.

Page 238, *Those who are most vulnerable often succumb to these secondary disasters:* John T. Watson, Michelle Gayer, and Maire A. Connolly, "Epidemics after Natural Disasters," *Emerging Infectious Diseases* 13, no. 1 (2007), doi:10.3201/eid1301.060779, https://wwwnc.cdc.gov/eid/article/13/1/06-0779_article.

Page 239, *All of this is why climate change has been called "slow violence":* Rob Nixon, *Slow Violence and the Environmentalism of the Poor* (Cambridge, MA: Harvard University Press, 2013).

Page 239, *The World Bank reports that disasters like floods:* Daniel Graeber, "World Bank to Invest up to $3.5B Annually for Renewable Energy Programs," *Albawaba*, April 11, 2016, https://www.albawaba.com/business/world-bank-invest-35b-annually -renewable-energy-programs-827554.

Page 239, *tribal elders insisted that the campaign stay both prayerful and peaceful:* Bill McKibben, "Indigenous Activists at Standing Rock Told a Deep, True Story," *The Nation*, December 5, 2016, https://www.thenation.com/article/indigenous -activists-at-standing-rock-told-a-deep-true-story.

96. Run for Office

Page 242, *Calling his the "Best Party," he publicized a platform with inane promises:* Jón Gnarr, *GNARR! How I Became the Mayor of a Large City in Iceland and Changed the World* (Brooklyn, NY: Melville House Publishing, 2015), 75.

Page 242, *upping his promises to include "a drug-free Parliament by 2020":* Sally Mc-Grane, "Icelander's Campaign Is a Joke, Until He's Elected," *New York Times*, June 25, 2010, http://www.nytimes.com/2010/06/26/world/europe/26 iceland.html.

Page 242, *In an email to supporters, Anderson wrote: "We've watched":* Erin Hegarty, "Councilwoman Becky Anderson Latest Dem to Announce Bid for Roskam's Seat," *Naperville Sun*, July 27, 2017, http://www.chicagotribune.com/suburbs /naperville-sun/ct-nvs-becky-anderson-congress-st-0728-20170727-story.html. See also Claire Kirch, "Indie Bookseller Becky Anderson Running for Congress," *Publishers Weekly*, July 28, 2017, https://www.publishersweekly.com/pw /by-topic/industry-news/bookselling/article/74347-indie-bookseller-becky -anderson-running-for-u-s-congress.html.

Page 243, *it might help to remember one of Gnarr's motivations: He was unemployed:* Gnarr, *GNARR!*, 65.

97. Sue the Grown-Ups!

Page 245, *footage of the children holding up their hand-drawn posters:* PBS NewsHour, "Teens Sue Oregon to Stop Climate Change," April 20, 2015, http://www.pbs.org/newshour/extra/daily-videos/teens-sue-oregon-to-stop-climate-change.

Page 246, *That time, Judge Ann Aiken affirmed Coffin's decision:* Juliana v. United States, 6:15-cv-01517-TC (Oregon 2016), accessed October 9, 2017, https://static1.squarespace.com/static/571d109b04426270152febe0/t/5824e85e6a49638292dddic9/1478813795912/Order+MTD.Aiken.pdf.

Page 247, *and they continue to work hard on cases pending in other states:* "State Judicial Actions Now Pending," Our Children's Trust, accessed August 21, 2017, https://www.ourchildrenstrust.org/pending-state-actions.

Page 247, *At this writing, children in twelve countries:* "Global Legal Actions," Our Children's Trust, accessed August 21, 2017, https://www.ourchildrenstrust.org/global-legal-actions.

Page 245, *Big Oil tiptoed out the back:* John Schwartz, "Students, Cities and States Take the Climate Fight to Court," *The New York Times*, August 10, 2017, https://www.nytimes.com/2017/08/10/climate/climate-change-lawsuits-courts.html?_r=1.

Page 245, *The Trump administration, meanwhile, is fighting the children:* Chelsea Harvey, "An Appeals Court Just Pressed Pause on the Much-Watched Youth Climate Lawsuit Against Trump," *The Washington Post*, July 27, 2017, https://www.washingtonpost.com/news/energy-environment/wp/2017/07/27/an-appeals-court-just-pressed-pause-on-the-much-watched-youth-climate-lawsuit-against-trump/?utm_term=.450f96d5c409.

98. Light Yourself on Fire

Page 248, *In 2010, a Tunisian street vendor protested government oppression:* Kareem Fahim, "Slap to a Man's Pride Set Off Tumult in Tunisia," *New York Times*, January 21, 2011, http://www.nytimes.com/2011/01/22/world/africa/22sidi.html.

99. Okay, Okay, Shrink Your Carbon Footprint

Page 250, *consider all the other everyday actions you might take from this list of possibilities:* This chapter's list of actions was inspired by and adapted ideas from the following sources: Chris Goodall, "How to Reduce Your Carbon Footprint," *The Guardian*, January 19, 2017, https://www.theguardian.com/environment/2017/jan/19/how-to-reduce-carbon-footprint; "Ten Things You Can Do to Shrink Your Carbon Footprint," *The Nation* (article conceived by Walter Mosley, with research by Rae Gomes), February 25, 2010, https://www.thenation.com/article/ten-things-you-can-do-shrink-your-carbon-footprint; "25+ Ways to Reduce Your Carbon Footprint," Carbon Offsets to Alleviate Poverty, accessed October 31, 2017, https://cotap.org/reduce-carbon-footprint; and "Energy Saver Guide: Tips on Saving Money and Energy at Home," US Department of Energy, accessed

October 31, 2017, https://energy.gov/energysaver/services/energy-saver -guide-tips-saving-money-and-energy-home.

Page 252, *Recycle motor oil. Nasty stuff, but it can be reused:* David Biello, "Can Oil Be Recycled?" *Scientific American*, August 25, 2009, https://www.scientificamerican .com/article/can-oil-be-recycled.

Page 252, *Fly economy... Prince William does:* Genevieve Shaw Brown, "Prince William Flies Coach on American," *ABC News*, May 7, 2014, http://abcnews.go.com /blogs/lifestyle/2014/05/prince-william-flies-coach-on-american.

Page 252, *Buy organic brews (commercial hops are liberally sprayed):* David H. Gent et al., eds., *Field Guide for Integrated Pest Management in Hops* (Moxee, WA: Washington Hop Commission, 2009), https://www.ars.usda.gov/ARSUserFiles/37109/Hop Handbook2010.pdf.

Appendix D: Kathleen Dean Moore's Letter to Grandparents

Page 263, *This letter by Kathleen Dean Moore appears in:* Kathleen Dean Moore and Michael P. Nelson, eds., *Moral Ground: Ethical Action for a Planet in Peril* (San Antonio, TX: Trinity University Press, 2010), 65.

Appendix E: Preamble to the US Constitution

Page 263, *According to the Cornell Legal Information Institute, the Preamble:* "U.S. Constitution: Preamble," Legal Information Institute, Cornell Law School website, accessed October 30, 2017, https://www.law.cornell.edu/constitution/preamble.

Index

About the Author

Mary DeMocker uses the arts to fight for a livable planet. Before turning her creative efforts to climate justice, she dressed sets for New York film productions, taught classical and folk music for three decades, and wrote children's plays that are now performed regularly in schools in the United States and abroad.

In 2013, Mary cofounded and became creative director of 350 Eugene, which plays a key role in Northwest climate justice efforts. She leads public art projects and climate justice events, including one featured in a *PBS NewsHour* broadcast about children suing the government for their right to a livable planet. She lectures regularly on re-storying the future and creative disruption, the topic of her interview in the Swedish documentary and multimedia installation *BiFrost*. In conjunction with the 2015 Paris climate talks, Mary designed and co-led the youth-centered Global Climate March & Collaborative Art Project that was featured in the London-based global climate art festival ArtCOP21 and included in an Avaaz video shown to world leaders as they entered UN climate talks.

A National Endowment for the Arts grant recipient and winner of the 2008 Kay Snow Award for Nonfiction, Mary has published in *The Sun*, *EcoWatch*, *Common Dreams*, Mothering.com, *Spirituality & Health*, and *ISLE* (*Interdisciplinary Studies in Literature and Environment*). She lives with her husband, children, and eleven chickens in Eugene, Oregon. Mary is available for workshops and presentations and can be reached through her website at www.marydemocker.com.